Artificial Intellig[ence]
with Power BI

Take your data analytics skills to the next level by leveraging the AI capabilities in Power BI

Mary-Jo Diepeveen

BIRMINGHAM—MUMBAI

Artificial Intelligence with Power BI

Publishing Product Manager: Reshma Raman
Senior Editor: Tazeen Shaikh
Content Development Editor: Sean Lobo
Technical Editor: Devanshi Ayare
Copy Editor: Safis Editing
Project Coordinator: Aparna Ravikumar Nair
Proofreader: Safis Editing
Indexer: Tejal Daruwale Soni
Production Designer: Shankar Kalbhor
Marketing Coordinator: Priyanka Mhatre

First published: April 2022

Production reference: 1240322

Published by Packt Publishing Ltd.
Livery Place
35 Livery Street
Birmingham
B3 2PB, UK.

ISBN 978-1-80181-463-8

www.packt.com

Contributors

About the author

Mary-Jo Diepeveen is an AI expert at Microsoft who focuses primarily on creating educational content around data and AI. Mary-Jo's passion for AI stems from her experience in neuroscience, where collecting and analyzing brain data to understand unconscious behavioral patterns led to an interest in machine learning and AI. This interest has continued to grow during her years at Microsoft, and she is often invited to talk and inspire others to get involved in tech.

About the reviewers

Divya Sardana serves as the lead AI/ML engineer at Nike. Previously, she was a senior data scientist at Teradata Corp. She holds a Ph.D. in computer science from the University of Cincinnati, OH. She has experience working on end-to-end ML and deep learning problems involving techniques such as regression and classification. She has further experience in moving the developed models to production and ensuring scalability. Her interests include solving complex big data and ML/deep learning problems in real-world domains. She is actively involved in the peer review of journals and books in the area of ML. She has also served as a session chair at ML conferences such as ICMLA 2021 and BDA 2021.

Art Tennick is the author of 20 computer books and over 1,000 magazine and LinkedIn articles, including one of the first books on DAX, written in 2009. He is an independent freelance consultant on Power BI (since 2009), Analysis Services (since 1998), and SQL Server (since 1993). His main interests are in Analysis Services (on-premises, Azure, and in the Power BI service), Power BI, Power BI Paginated, and the integration of Python and R data science with Power BI, Analysis Services, and SQL Server. You can find him on LinkedIn.

I would like to thank Rita Mendoza, Emma, and Lorna for their love and support over the years.

Ricardo Franco da Silva has worked at Microsoft as a consultant for the BI stack and Azure for almost 20 years. Ricardo has hasparticipated in projects for sectors including oil, banking, aviation, retail, and governments worldwide, managing big data and getting insights using many tools, especially SQL Server, Analysis Services, and Power BI, among many others. One of the most challenging projects was for a bank in Switzerland. Ricardo cooperated with Packt to review this book, and he also wrote a novella a few years ago. With over 30 years of experience, Ricardo is currently engaged in projects related to AI and recently concluded a course in fintech AI and open banking at the University of Oxford.

Table of Contents

3

Data Preparation

Part 2: Out-of-the-Box AI Features

4

Forecasting Time-Series Data

5

Detecting Anomalies in Your Data Using Power BI

6

Using Natural Language to Explore Data with the Q&A Visual

7

Using Cognitive Services

8
Integrating Natural Language Understanding with Power BI

9
Integrating an Interactive Question and Answering App into Power BI

10
Getting Insights from Images with Computer Vision

Part 3: Create Your Own Models

11

Using Automated Machine Learning with Azure and Power BI

12

Training a Model with Azure Machine Learning

13

Responsible AI

Index

Other Books You May Enjoy

Preface

Even though many organizations are embracing **Artificial Intelligence (AI)**, the field remains mystical for many. AI is a large field with various applications, which makes it difficult to completely comprehend if you are new to the field.

The focus of AI is often to find the best-performing model to predict new data. However, equally, if not more, important is understanding how to apply those models to ensure the predictions are properly consumed by the targeted users.

Power BI is a great tool for extracting the value out of AI models. Data analysts are already familiar with working with data and are the ideal candidates to integrate a model's predictions into a story to tell decision-makers.

With this book, I hope to demystify AI and make it more approachable for data analysts working with Power BI. The aim of this book is not to become a full-fledged AI expert or data scientist, but to understand what a data scientist deals with to create models. Based on the fundamental understanding of AI, you should be able to consciously decide whether AI will enrich your reports in Power BI.

Since I wrote this book to show how AI can be applied to create value, there is a large focus on walkthroughs that you can replicate. If you prefer to learn by doing, and are keen to explore how to apply AI, this book is for you.

Who this book is for

The book focuses on applying AI in Power BI, which means some experience or familiarity with Power BI is preferred. The AI features will be introduced assuming you have no prior knowledge. Most of the features are accompanied by a walkthrough, to help you understand the concepts that are explained in each chapter.

Although this book will introduce you to many AI concepts, it is not meant to teach you how to do data science. Feel free to use this as a stepping board but know that there is much more to learn in order to become a data scientist.

What this book covers

Chapter 1, Introducing AI in Power BI, introduces basic concepts associated with AI and Power BI.

Chapter 2, Exploring Data in Power BI, introduces some Power BI features to explore data as is often done in data science.

Chapter 3, Data Preparation, introduces some data requirements when implementing AI and how those requirements can be met within Power BI.

Chapter 4, Forecasting Time-Series Data, introduces forecasting as a data science method and how to use the built-in forecasting feature in Power BI.

Chapter 5, Detecting Anomalies in Your Data Using Power BI, introduces anomaly detection as a data science method and how to use the built-in anomaly detection feature in Power BI.

Chapter 6, Using Natural Language to Explore Data with the Q&A Visual, introduces natural language querying and how to use the built-in Q&A visual in Power BI.

Chapter 7, Using Cognitive Services, goes through the various AI models offered through Azure Cognitive Services and how they can be of use.

Chapter 8, Integrating Natural Language Understanding with Power BI, further explores Cognitive Services for Language and how to apply its models in Power BI.

Chapter 9, Integrating an Interactive Question and Answering App into Power BI, focuses on one of the Cognitive Services for Language models and how you can create a power app to create an interactive Q&A visual in Power BI.

Chapter 10, Getting Insights from Images with Computer Vision, further explores Cognitive Services for Vision and how to apply its models in Power BI.

Chapter 11, Using Automated Machine Learning with Azure and Power BI, introduces Azure Machine Learning and how to use its automated machine learning feature to quickly train multiple models. The best-performing model will be integrated with Power BI.

Chapter 12, Training a Model with Azure Machine Learning, covers how to train your own model in Azure Machine Learning, and how to integrate a model with Power BI.

Chapter 13, Responsible AI, discusses important considerations when working with AI to ensure its fair and responsible use.

To get the most out of this book

All walkthroughs will require Power BI Desktop installed on a computer. Next to that, you'll need an Azure subscription to access the Azure portal through a browser.

Software/hardware covered in the book	Operating system requirements
Power BI Desktop	Windows
Azure subscription, preferably associated with a Microsoft account	

Not all chapters' walkthroughs require an Azure subscription. You'll be informed in each chapter when you need access to Azure and instructions on how to use it are included.

If you are using the digital version of this book, we advise you to type the code yourself or access the code from the book's GitHub repository (a link is available in the next section). Doing so will help you avoid any potential errors related to the copying and pasting of code

Download the example code files

You can download the example code files for this book from GitHub at `https://github.com/PacktPublishing/Artificial-Intelligence-with-Power-BI`. If there's an update to the code, it will be updated in the GitHub repository.

We also have other code bundles from our rich catalog of books and videos available at `https://github.com/PacktPublishing/`. Check them out!

Download the color images

We also provide a PDF file that has color images of the screenshots and diagrams used in this book. You can download it here: `https://static.packt-cdn.com/downloads/9781801814638_ColorImages.pdf`.

Conventions used

There are a number of text conventions used throughout this book.

`Code in text`: Indicates code words in text, database table names, folder names, filenames, file extensions, pathnames, dummy URLs, user input, and Twitter handles. Here is an example: "If we look at the `Country name`, we can see we have 149 distinct and 7 unique values."

A block of code is set as follows:

```
import matplotlib
dataset.boxplot("Life Ladder", showmeans=True,
showfliers=False)
plt.show()
```

Any command-line input or output is written as follows:

```
pip install pandas
pip install matplotlib
```

Bold: Indicates a new term, an important word, or words that you see onscreen. For instance, words in menus or dialog boxes appear in **bold**. Here is an example: "Select **Transform Data**, this will open the **Power Query Editor**."

> **Tips or Important Notes**
> Appear like this.

Get in touch

Feedback from our readers is always welcome.

General feedback: If you have questions about any aspect of this book, email us at customercare@packtpub.com and mention the book title in the subject of your message.

Errata: Although we have taken every care to ensure the accuracy of our content, mistakes do happen. If you have found a mistake in this book, we would be grateful if you would report this to us. Please visit www.packtpub.com/support/errata and fill in the form.

Piracy: If you come across any illegal copies of our works in any form on the internet, we would be grateful if you would provide us with the location address or website name. Please contact us at copyright@packt.com with a link to the material.

If you are interested in becoming an author: If there is a topic that you have expertise in and you are interested in either writing or contributing to a book, please visit authors.packtpub.com.

Share Your Thoughts

Once you've read *Artificial Intelligence with Power BI*, we'd love to hear your thoughts! Scan the QR code below to go straight to the Amazon review page for this book and share your feedback.

https://packt.link/r/1-801-81463-5

Your review is important to us and the tech community and will help us make sure we're delivering excellent quality content.

Part 1: AI Fundamentals

This part of the book will help you understand what AI can do and how it can be used in Power BI. It will show you how to make sure your data is prepared and that you understand which AI feature to use.

This section includes the following chapters:

1
Introducing AI in Power BI

Everyone wants to be working with data. Organizations are keen on relying more on data-driven decisions instead of intuition-driven decisions. To be driven by data, we need to extract insights from data. Thankfully, **Power BI** is a great tool to visualize and share what the data tells us. To better understand what trends and learning we can derive from data, we can use techniques from the field of data science.

Data science and **artificial intelligence** (**AI**) are becoming increasingly popular approaches to extracting insights from data. Among other things, this is because these tools allow us to work with unstructured data, which we couldn't work with before. And this helps us to more quickly find complicated trends and patterns in the data.

In this book, we will focus on using Microsoft Power BI as a data exploration and visualization tool. And we will take some parts of the **Azure cloud** to give us the power to train models and integrate this with Power BI.

But first, let's start with some of the groundwork. We need to understand what AI is, to properly scope our projects and run them successfully. We need to know what is possible and how we go from a simple dataset to an AI model before we get into the details of each step of the process. That is why we will first dive into the following questions:

- What do we expect from a data analyst?
- What is AI?
- Why should we use AI in Power BI?
- What are our options for AI in Power BI?

Let's start with covering the basics.

What do we expect from a data analyst?

Every company is looking for different insights and is working with different types and sets of data. Even though you will find data analysts across several organizations, their actual day-to-day work can greatly differ. When reading this book, you will pick up whatever is useful to you, and you will probably skip over irrelevant parts. Nevertheless, it is good to go over what we expect you to know and be familiar with.

First, we will discuss what it means to be a data analyst, the assumptions we make, and why we chose to use this job title. Next, we will go over what you should already know about Power BI and where you can find this information if you feel you need to brush up on that knowledge.

What is a data analyst?

You could call yourself a business intelligence engineer, a business intelligence specialist, a database administrator, or simply a data analyst. Whatever your job title is, you picked up this book because you work with Power BI and want to learn more about it. With all these different titles nowadays, it becomes challenging to understand what your base knowledge should be. For simplicity and consistency, we will refer to a person working with Power BI as a data analyst.

Why a data analyst? Because in this book, we will assume that you are familiar with working with data in Power BI and are able to do the following:

- Prepare data
- Model data (creating a data model in Power BI, not a machine learning model)

- Visualize data

- Analyze data

- Deploy and maintain Power BI deliverables

On the other hand, we'll assume you're unfamiliar with the field of data science. We'll approach all AI features in Power BI from the point of view of this persona. In this book, we'll introduce the machine learning models behind the AI features, to make sure enough is understood to use the features correctly. However, we won't go into the full complexities of all models, as this is not a book targeted at the data scientist, who already has a lot of knowledge about mathematical and statistical methods used in AI.

There are two main skills important for when you are venturing into AI in Power BI: connecting to and visualizing data. Let's elaborate a little bit more on these two topics so that you know what is expected of you before you continue.

Connecting to data

So, assuming we are all data analysts, let's look at our core tasks. The very first thing we need to do to work with data, is *get access to data*. From a technical perspective, we can very easily connect Power BI to various data sources, whether we have data stored in cloud databases, Azure or otherwise, or whether we have local files we want to connect to. Power BI will allow us to do so and will even allow us to schedule an automatic refresh to visualize new data as long as we set up a gateway connection between the network within which the data resides and the Power BI service.

What kind of data can you work with? Any kind! You can connect to *structured* data, formatted nicely in tables, *semi-structured* data, often in the form of JSON, or even *unstructured* data to insert images into your Power BI reports. This also means data can come from a variety of sources. You can collect Twitter data (semi-structured) which contains tweet text, date of creation, number of retweets, likes, and hashtags used. You can collect sales and marketing data to understand which products you have sold, when you sold them, and which ad campaigns you were running that may have had an effect on your sales. Or maybe you are looking at the supply and demand for your warehouses and stores to make sure you plan the logistics of stocking your stores accordingly.

Since data can be generated by so many different sources, and can come in so many different formats, we also want to think about how we *extract that data* and get it ready to build reports on. Power BI has a lot of standard connectors to allow you to connect to data. A best practice here, however, is that you have a pipeline handling data orchestration before you even connect Power BI to it. Such a process is often called an **ETL (Extract-Transform-Load)** or **ELT (Extract-Load-Transform)** pipeline in which we connect to our sources generating data, extract the data, load it into a database, and transform it if necessary. Although similar tasks can be done by Power BI, we prefer working with ETL tools such as **Azure Data Factory** to handle these kind of orchestration pipelines when we are working with large amounts of data.

> **ETL or ELT?**
> ETL is an established and commonly used approach to extract data from sources. The purpose is often to transform it and load it into a structured database such as an **Azure SQL Database** and fit the data nicely into columns and rows. This is very well suited for transactional data, where we want to get quick results. However, with the emergence of cloud, the newer approach, ETL, is getting more traction. By extracting and loading the data into a data storage that can store unstructured data, such as an **Azure Storage Account** or **Azure Data Lake**, we can dump data in the cloud without having to worry about schemas. This also makes it easier to reuse the same data and transform it in different ways, depending on the insights you want to extract.

Long story short, a lot can and probably should already happen to your data before you even open Power BI. Be aware of the process that may have occurred before data enters Power BI. Whatever source we use for Power BI will influence the options we have within Power BI, as well as the performance of the reports we create. In this book, we will work mostly with preprocessed data, available through files stored on public websites or cloud databases. For some projects with AI, however, we will need to process data before bringing it into Power BI to make sure we can work with it.

Visualizing data

Lastly, the reason we are using Power BI is because we want to *tell a story with our data*. We want to convert data into valuable and intuitive insights that everyone within our organization can read for their own use. In the context of AI, this may be one of the most important skills of a data analyst, as will be explained in later sections.

So, what do we mean by telling a story? Humans are much more susceptible to stories than listening to data. If we say there is a 70% chance it is going to rain, do you bring an umbrella? It's hard to decide. If someone tells you to bring an umbrella, you will probably do it. Even though we want to make more data-driven decisions, humans are not naturally driven by data. We are driven by stories, which we find more intuitive. That also means that we cannot just give AI insights to humans. We need to *translate* the output of an AI model to make sure it is understandable for people. That means we need to use the data to tell stories.

The way we can do it within Power BI is by making use of the visuals it offers. We have the option to use standard visuals, import visuals from the marketplace, or create our own visuals using **Python** or **R**. Understanding when to use what and how to combine different visuals into a report is an important skill of a data analyst, and a skill we assume you have when reading this book. Throughout the book, remember that to gain people's trust, we need to speak their language and not just throw numbers at them to make them change their behavior.

In this book, we will focus on how we can use Power BI to perform AI. That means that you, as a data analyst, are already familiar with the different types of data you can get into Power BI from the different sources available to you. You should already be familiar with how to create visuals to form a report in Power BI. In later chapters, we will highlight features that are relevant when preparing your data for AI and to actually implement AI. But first, let's talk more about what AI is.

What is AI?

AI is a term often used to show that organizations use state-of-the-art technologies. Interestingly, this term has already existed for over 60 years. Back then, it was defined as *the science and engineering of making intelligent machines* (Professor John McCarthy, Stanford University, accessed June 2021, `http://jmc.stanford.edu/artificial-intelligence/what-is-ai/`). Unfortunately for us, that leaves a lot open for interpretation, which is also why the term *AI* has had so many different meanings over the years.

AI often goes hand in hand with data science, which is a field combining science, engineering, mathematics, and statistics. Its purpose is very much in the name: it's the science of extracting insights from data, to make sense out of the very raw data we might get from any applications or databases we have. Using this field, we can get data, clean it up, train a model based on that data, and integrate that model into our applications to generate predictions on new incoming data.

To fully grasp what AI can do, we need to understand a couple of different terms, often used together with AI: **machine learning**, **deep learning**, **supervised learning**, and **unsupervised learning**. Next to that, it helps to be familiar with a typical structure of the process it takes to create an AI model.

Understanding the definition of AI

We can find many different definitions for AI, which generally have three main aspects in common:

- Computers
- Executing an intelligent task
- Like a human would do it

Computers come in different forms and can mean software or hardware; it can be an application running locally on someone's laptop or an actual robot. The part of the definition of AI which is more open to interpretation is the intelligent task executed like a human would. What do we consider *intelligent*? If we think about an intelligent task performed by a human, we could also take the example of a calculator. The more expensive calculators are able to make complex calculations within seconds, which would take a mathematician some time to figure out. However, if you asked someone whether a calculator should be considered AI, the answer would most probably be *no*.

So, then the question arises: what is *intelligence*? Fortunately, there are many philosophers who are spending their academic life on answering this question, so let's assume that is outside the scope of this book. Instead, let's agree that the threshold of what is considered to be AI evolves over the years. With new developments come new expectations. Whereas we first considered beating the world champion in chess to be the ultimate level of AI, we now wonder whether we can create fully autonomous self-driving cars.

Some of these new developments have been new types of algorithms that allow us to train even more intelligent models. These algorithms are often categorized as *machine learning* and *deep learning* algorithms and are important to understand to know when to use what. We will explore both of these terms in more detail.

Understanding machine learning

If we think back to our simple yet high-performing calculator, you can imagine the intelligence of such a machine being created by a rule-based system. Adding one plus one is always two. This kind of mathematical rule and many others can be programmed into a calculator to empower it to count. This approach is also called **using regular expressions** and can still be very useful today. It is considered the most rudimentary approach to accomplishing AI but can still yield quick and clear results.

If you want smarter AI, however, you might want to work with techniques where a model is not fully programmed based on rules we humans decide on, but instead is **self-learning**. Because of this term, it is often thought AI is **self-improving** and will continuously improve over time until it reaches the **singularity**. What self-learning actually means is that we do not have to explicitly tell AI how to interpret data coming in. Instead, we show AI a lot of examples and, based on those examples, the model we train will decide how a pattern of variable values influence a specific prediction.

For example, what if you sell laptops and you want to advertise the right laptop to the right person? You could work with a rule-based system where you would create groups based on demographic data, such as women younger than 30, women who are 30 or older, and the same for men younger than 30 and men who are 30 or over. We would have four different groups we would use different marketing strategies on, assuming that every woman younger than 30 has the same requirements when buying a new laptop.

Instead of this, we of course want to pick up on patterns we may not have realized ourselves but can still be found in the data. That is when we would use machine learning to have a self-learning model that looks at the data and learns which variables or features make you interested in specific laptop requirements. It could very well be that, based on this self-learning, we find out that we have different groups we should use different marketing strategies on. For example, we may have a subgroup of both men and women under 30 who love to play online games and need different requirements than men and women under 30 who only use their laptop for work.

Compared to regular expressions, using machine learning to accomplish AI is considered a more sophisticated approach. To create a machine learning model, we take our **data**, and we choose an **algorithm** to **train a model**. These terms are visualized in *Figure 1.1*:

Figure 1.1 – Creating a machine learning model

As seen in *Figure 1.1*, the data you have and the algorithm you choose are the inputs, and the model you train is the output. The algorithm we select will decide how we want to look at our data. Do we want to classify our data? Do we want to create a regression model where we want to predict a numeric value? Or do we want to cluster our data into different groups? This information is encapsulated in the algorithm you choose when training a model.

Now that we understand what machine learning is, how does it then differ from deep learning?

Understanding deep learning

Machine learning has already opened up a world of possibilities for data scientists. Instead of spending hours and hours exploring data and calculating **correlations**, **covariances**, and other statistical metrics to find patterns in the data, we could just train a model to find that pattern for us.

Machine learning was initially done with structured data fitting nicely into columns and rows. Soon, people wanted more. Data scientists wanted to also be able to classify images or to understand patterns in large text documents. Unfortunately, the algorithms within machine learning could not handle these kinds of unstructured data very well, mostly because of the complexity of the data itself. It is said that an image says a thousand words and it is indeed true that even one single pixel of an image holds a lot of information that can be analyzed in many different ways.

With the emergence of **cloud computing** and the improvements in processing units, a new subfield within machine learning arrived. Instead of the simpler **CPUs** (**Central Processing Units**), we now have the more powerful **GPUs** (**Graphical Processing Units**) that can process complex data such as images at a much faster rate. With more power comes more cost, but thanks to the cloud, we have GPUs available on demand and only have to pay for when we use them.

Once we had the processing power, we still needed different algorithms to extract the patterns in these kinds of unstructured data. Since we wanted to perform these tasks just like humans would do it, researchers turned to the brain and looked at how the brain processes information. Our brain is made up of cells that we call *neurons*, which process information on many different layers. So, when looking at unstructured data, researchers tried to recreate a simplified **artificial neural network** in which these neurons and layers are simulated. This turned out to work very well and resulted in the subfield of deep learning. Now, we can take images and classify them or detect objects in them using the subfield we call **Computer Vision**. And we can use **Natural Language Processing** (**NLP**) to extract insights from text.

We have now talked about AI, machine learning, and deep learning. In *Figure 1.2*, a visual overview is shown of these three terms:

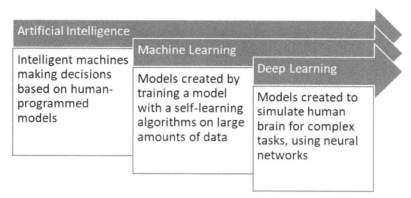

Figure 1.2 – Overview of AI, machine learning, and deep learning

As seen in *Figure 1.2*, the three terms we just discussed are related to each other. AI is often considered to be the umbrella term for anything we do to let a machine execute an intelligent task like a human would do it. One approach to this has been machine learning, in which we train models not by teaching them the rules, but by letting the models learn themselves. All we do is offer the data and algorithm (which defines the task) to train a model. And lastly, we have deep learning as a subfield of machine learning, in which special algorithms are used to better handle complex tasks such as understanding patterns in unstructured data.

Besides these three main terms that are necessary to understand before working with AI in Power BI, we also need to make a distinction between *supervised* and *unsupervised* learning. These two approaches divide the algorithms we will use into two categories. Understanding the difference between them helps us to know what we need to include in our datasets that we will use as input for a model we want to train.

Understanding supervised and unsupervised learning

The purpose of AI is that you want to predict something. You want to predict something such as whether someone is more likely to buy a washing machine or a new fridge. Or you want to predict how many apples you will sell on a given day so that you know how much to supply your store with. What we want to predict is often called a **label** or **tag**. Sometimes we have training data that includes that label, and sometimes we do not.

If we run a store that sells apples, we can take our historical data, which we can combine with extra data such as weather data and day of the week. On cold Mondays, no one may want to buy apples but on sunny Fridays, you may run out of apples before noon. Because we can see in our historical data how many apples we have sold in the past when specific conditions were met, we have training data that includes a label, namely number of apples sold. If we know the label, we call this *supervised learning*.

What about the laptops we were selling? Let's say we have customer data including demographic information such as age and gender. But we may also have data on what they use the laptops for: online games or work. In this case, we do not know how many different groups we should create different marketing strategies for. So, the groups we want to categorize them into do not exist in the training data. Therefore, we have no labels in our training data and thus are doing *unsupervised learning*.

It is good to make the distinction between these different terms because it will help you understand what is required from your data and what you can expect from the model. Finally, let's zoom in to the different types of algorithms that we can expect to use throughout this book.

Understanding algorithms

When you have decided you want to apply AI to your data, we now know that we need to have data and an algorithm to create a model. We will discuss the requirements of the data in later chapters in much more detail. That leaves us with understanding how to use algorithms. Understanding how to work with algorithms is considered to be the data scientist's expertise as it lies at the cross-section of mathematics and statistics. However, even if we do not build models ourselves, or aspire to becoming full-on data scientists, it is still beneficial to understand the main types of algorithms we can work with.

The most important thing we need to know is that by choosing the right algorithm, we dictate how we want to look at the data and what kind of pattern should be detected by the model. If we talk about supervised learning – where we know the label we want to predict – we often talk about **regression** or **classification**. With regression, we try to predict a numeric value, whereas with classification the label is categorical (two or more categories).

Understanding regression algorithms

Imagine you work for a company that supplies electricity to households all across the country. You will have collected a lot of data on your customers, such as where they live, what type of house they live in, the number of people who make up that household, and the size of the house. For existing customers, you know how much energy they have consumed in previous years. For new customers, you want to predict what their energy consumption will be so that you can make a good estimation on what the costs will be for them.

Our data may look something like the table shown in *Figure 1.3*, where we have historical data for two existing customers, and we want to predict the energy consumption in kWh for the third, new customer:

Location	Type of house	Number of people	Size of house (m²)	Energy consumption (kWh)
City	Apartment	2	50	1990
Town	House	4	120	4320
City	Apartment	3	70	???

Figure 1.3 – Data for three customers on household characteristics as well as energy consumption

In this example, we know the label: energy consumption. So, we know that we are doing supervised learning. The variable we want to predict has a *numerical value* (for example 1990 kWh, 4320 kWh, or anything in between). Therefore, this is a simple example of a regression model. There are different algorithms we could choose within the subset of regression to train such a model. The choice may depend on things such as how complicated you want to allow your model to be, how explainable you want your model to be, and how much compute and time you want to spend on training your model. Some examples of such algorithms are **linear regression**, **decision forest regression**, and **boosted decision tree regression**.

Understanding classification algorithms

After we trained our model with one of these algorithms and our historical data, we were able to correctly predict for our new customer what the energy consumption will be and how much they will be spending per month. Our new customer has agreed to buy electricity from us but wants to know more about potential ways to save energy. This brings us to the idea of working more proactively with solar panels. If people have solar panels on their roofs, they can generate electricity themselves on sunny days and save money. We, of course, want to help them with this investment and the installation of solar panels.

Some customers may already have solar panels, some may approach us to talk about solar panels, and some may have never given it a thought. We want to reach out to customers and advertise the solar panels that we sell, but we do not want to annoy or spend marketing budget on customers who already have them.

So, what we want to do now is make an inventory of which of our customers already has solar panels. Sure, we could check with each and every household to see whether they have solar panels, but that seems like a big task that takes too much time and energy. And not every household may react to the survey we would send out to collect that data. We therefore decide we are going to see whether we can predict it. We can collect some sample data, train a model that can predict whether that customer has solar panels, and use those insights to target the right households.

This sample data may look like the data shown in *Figure 1.4*. We again have historical or known data, and we have a label. Since we know the label for our sample data, we are doing supervised learning. In this case, however, we are not trying to predict a numeric value. The label we try to predict is *solar panels*, and can either be *yes* or *no*. So, we have two categories, making it a classification problem; more specifically, a *binary* or *two-class* classification problem:

Location	Type of house	Number of people	Size of house (m²)	Energy consumption (kWh)	Solar panels
City	Apartment	2	50	1990	No
Town	House	4	120	4320	Yes
City	Apartment	3	70	2490	???

Figure 1.4 – Household characteristics and the solar panels label

Again, there are different algorithms we could choose from when we know we want to do classification. First of all, it matters whether we have two or three or more classes for our label. Next, we again can choose how complex or explainable we want our model to be with options such as a **two-class logistic regression** or **decision forest** algorithm to train our model with.

Finally, let's take look at one more simple example so that we have seen different types of algorithms we can work with. Imagine we are still supplying electricity to people all across the country, but now a new competitor has entered the market. We are afraid this new business will take away some of our customers, so we want to offer people nice benefits to convince them to stay with us. We do not want to offer this to all our customers, to save money, meaning we need to make a good assessment on who will leave us.

We have not had many customers leaving us yet, and we want to prevent this from happening. This does mean, however, that we do not have a label; we do not have enough data on the variable we want to predict. In this case, we might want to cluster our customers, and divide them into different groups: those who will leave us for the competitor, and those who will not. We may have some customers that have left us. Based on the little data we have, we can use an algorithm such as **K-means clustering** to find similar data points to the customers who left us to create these groups and target those that are on the verge of leaving us with attractive deals to make sure they stay with us.

Working with algorithms requires an understanding of the mathematics and statistics behind them. To use them to train models, we rely on data scientists bringing that knowledge to our team so that we can make more data-driven decisions. To work with the AI features in Power BI, we are not expected to become data science experts. It does, however, help to understand the choices being made to realize what the potential and restrictions are of using AI on our data.

Now that we know what AI is and how we can use different algorithms to train our model, let's take a step back and have a look at the complete **data science process**. How do we take data from beginning to end? And how we can we do so successfully?

What is the data science process?

Just like every project, training a model consists of multiple phases. And just like many projects, these phases are *not necessarily linear*. Instead, we want to take an iterative approach when developing an AI solution. First, let´s have a look at what the phases of the data science process look like.

The very first thing we need to think about is what we are doing this for. Why do we want to use AI? What is the model going to do? Even though it is good to drive innovation, we want to avoid using AI just because everyone else is doing it. Nevertheless, identifying an appropriate use case for AI can be challenging as many applications are relatively new and unknown. So, what is then a good use case? Of course, it depends on the area of business, but in each area, there is some low-hanging fruit that we can identify. Most commonly, we can think of using AI to predict how to invest marketing budgets, how to increase sales, how to monitor for predictive maintenance, or how to find outliers and anomalies in your data such as risk assessments.

After deciding on the use case, and essentially the scope, it becomes easier to think about what data to use and what metrics to evaluate the model on to know whether it will be successful or not. The next step is then to actually *get the data*. From a technical perspective, this could mean collecting data, building a new data orchestration pipeline to continuously extract data from a source such as a website or CRM system, or simply getting access to a database that already exists within the organization. Some hurdles may arise here. To train a good model, we need good data. And our data may not meet the requirements of having the right quantity or quality. There may also be **Personally Identifiable Information** (**PII**) data that needs to be masked or excluded first before we are allowed to work with that data.

Assuming that data is acquired by the data scientist, they can finally use their expertise to build the model. To build a model, we need data and an algorithm. The data we have received may need some processing. We may want to check for biases in the data, impute missing values, or transform data to make it more useful for our model. This phase is called **pre-processing** or **feature engineering**. The purpose is to end up with features that will serve as the input for our model.

Once we have a set of features, often in the form of variables structured as columns in a table, we can actually *train the model*. This means we try out algorithms and evaluate the models trained by looking at the resulting metrics provided by the different models. This phase by itself is very iterative and can require multiple models being trained (sometimes in parallel). After evaluating the model, based on the requirements from the use case, it can also lead to going back one or more phases to either redefine the use case, get different data, or alter the choices made in feature engineering.

Once it has been decided that a good enough model has been trained, the final phase can finally be executed. What exactly has to happen during this phase will depend on how the model's insights are being consumed. One example of how we can *integrate the model* is in a client application, where data is generated or collected in that application and is sent to the model to get real-time predictions back, which are also used in applications. Another common example is using a model for batch analysis of data. In this case, we can integrate the model into our data orchestration pipeline to make sure we use powerful compute to process a large amount of data. Whether it is real-time or batch predictions we want to generate, this is a final and crucial step we need to take into consideration when going through the data science process as shown in *Figure 1.5*:

Figure 1.5 – The five data science process phases

The data science process is not a linear process but understanding the five phases we most likely will iterate through can help us in knowing when to do what. A good project starts with a clearly defined use case, we then acquire data, prepare it through feature engineering, train a model with that data, and finally we integrate our model into our applications or Power BI reports.

Integrating AI with Power BI is of course especially interesting for the data analyst. In the next section, we will try to answer the question why this is a match made in heaven.

Why should we use AI in Power BI?

The question of why we should use AI in Power BI is twofold. First of all, we may wonder why we should use AI to begin with, and secondly, we may wonder why we should use the features in Power BI. To answer the former, we need an understanding of what AI can do, which is covered in earlier sections. To answer the latter, we need to understand why AI is not being adopted yet by most organizations.

The problems with implementing AI

There is an undeniably large interest in anything AI related. Unfortunately, like many new technologies, everyone loves to talk about it, but only few actually do it. There are many reasons why the adoption of AI is lower than expected (*McKinsey Survey on AI Adoption* from 2018, accessed June 2021, `https://www.mckinsey.com/featured-insights/artificial-intelligence/ai-adoption-advances-but-foundational-barriers-remain`). The most obvious one seems to be the *lack of skills*. In previous sections, we discussed what AI is and how we can create models. We discussed how, to train a model, we need to choose an algorithm and that this requires data science knowledge, which is, among other things, a combination of mathematics and statistics. Consequently, a large reason of why companies do not use AI is because they do not have employees with the expertise in building these models and understanding the math behind it.

This is not the biggest roadblock. Many software vendors already recognized this problem some time ago and created tools and services directed at the *citizen data scientist*, democratizing the technology and making it available for everyone who wants to use it, regardless of whether you have a degree in data science. These easy-to-use tools should not replace any AI investments but do raise the question of why AI is not being used more.

The answer may be because *people do not know about it*. This has less to do with actual data scientists being hired but more with the rest of the organization. At the higher levels, leadership does not know how to form clear strategies or a practical vision around AI. At the lower levels, employees do not know how to use AI in their day-to-day work and even when AI insights are provided for them to work with, they often end up not trusting the machine as opposed to their own intuition. It seems that for both leadership as well as employees, the problem lies with *an incomplete understanding of what AI can do* and recognizing the possibilities and restrictions of AI.

Even when companies have recognized the potential, created a clear strategy on implementing AI, and hired the right people for it, they face issues. When you ask a data scientist what the most challenging part of their job is, they rarely mention *training the model*. The most common hurdle revolves around the data. Either for political or technical reasons, *data scientists cannot get access to the right data*. The data they need might not even exist. And even when they do get access to the right data, it is often not of good quality or of enough quantity to train a good model with.

And finally, to really implement AI into your business processes, you not only need data scientists, but you also need a whole team. You need data engineers to help you with extracting data from their sources, clean the data at scale, and offer it to the data scientists who can then train a model. After a model is trained, you need to make sure the insights are offered to the business in an intuitive way, for which you will need software engineers to integrate the model into the client applications or data analysts to visualize the insights from the models in your Power BI reports.

In other words, to create an end-to-end solution and implement AI in an enterprise environment, you need an interdisciplinary approach, and *a collaboration between different departments*, where, preferably, you also want to ensure that everyone has a basic understanding of what AI can do to build trust and enhance adoption.

Why AI in Power BI is the solution

Now that we understand the problem, we can get an idea of what could be the solution. Since there are many reasons there is a slow adoption of AI, here is an overview:

- Lack of data science skills

- Incomplete understanding of AI

- Not enough, or not good enough, data

- No collaboration between departments

Unfortunately, finding skilled data scientists is very challenging. The alternative can be to train your employees, which is something that will also help against an incomplete understanding of AI that is seen across different layers of organizations. To collect better or more data, we need to know what we are doing it for. Why should we invest in this and what will be the benefit? And to stimulate collaboration between departments, we need to create understanding and trust.

One tool that can help with all of these things is Power BI. Compared to data scientists, companies have significantly more data analysts who are familiar with working with data and who are either already working with Power BI or will easily adapt to working with it. That means that data analysts already know the importance of good data and have access to data. Using Power BI, we try to tell a story with the insights we generate from data, to help people make data-driven decisions. Data analysts know how to convey numbers into intuitive facts. They can help convey AI output into information that can be understood and trusted by anyone within and outside the organization. This can consequently also help with facilitating collaboration between departments as Power BI can already be used across different departments.

The only blocker is that those using Power BI are often not familiar with AI. They might not have the data science expertise, but they are the ones who can work very well with them. By combining AI with Power BI, we can educate others, to help the business create clearer AI strategies and to help end users gain trust over the model's output and how it can help them in their business processes.

That is why this book covers the different AI options in Power BI. They include low-hanging fruit to get started with today, to show the possibilities of AI. But Power BI can also integrate with sophisticated models that have been trained by data scientists. It is therefore a logical starting point to adopt AI at a larger scale within your organization.

Now that we understand why it is so beneficial to use AI in Power BI, let's have a look at our options.

What are our options for AI in Power BI?

If we think about what we can do with AI in Power BI, there are roughly two categories we could put the options in. First of all, we have the low-hanging fruit: the easy projects with which we can start today. And second, we have the ability to create our own models and integrate these with our Power BI reports that give us more flexibility but require a larger time investment.

Out-of-the-box options

The easy AI options in Power BI can also be referred to as *out-of-the-box* AI features. These models are pre-built by Microsoft, meaning that we do not need to spend time on collecting data to train the model, nor do we need the expertise to choose the right algorithm. That already saves us time on the most challenging phases in the data science process!

For most of these features, the models are already integrated with Power BI and all we have to do is consume them. For others, we have the option to add a little bit of our own data to customize it to our business scenario. That means that there is some kind of base model under the hood that is already trained by Microsoft on data collected by Microsoft (check the *Privacy Agreement* for the service you are working with to see whether your data is being used). We then add our own data so that Microsoft can finish training a model in a fraction of the time it would take us to create the same model ourselves from scratch.

Next to that, we can also see these models being offered in various ways. We have integrated models in Power BI that are accessible through rich visuals, also known as AI visuals. We have integrated models that can be used with a specific type of data, such as the forecasting we can add when plotting time-series data. And finally, we can use the **Cognitive Services**, a collection of Azure services consisting of pre-built models that we can very easily integrate with any application through the use of APIs.

Creating your own models

The advantage of using pre-built models is that you save time and money when getting started with AI. The disadvantage of using these kinds of models from any software vendor is that you have less control and less flexibility over the design of your model. If, instead, you want to create your own models, you do need access to data science expertise to make the right decisions when training a model.

Still, there are many situations in which we want to make sure we create a model that is specifically designed for our use case. In this book, we'll assume we want to work with Microsoft's cloud platform Azure to easily integrate any machine learning model with Power BI. We have three main options in Azure when training our own model:

- Using **Automated Machine Learning** to train multiple models (in parallel), choose the best one, and integrate that one with our data pipeline.

- Using the **Azure Machine Learning Designer** to create a model.

  Both of these options require less data science expertise than the final option:

- Using the **Azure Machine Learning workspace** to train and deploy model based on training scripts that are created from scratch. The latter is what data scientists are often trained to do. They are comfortable in Python or R and most commonly use open source libraries such as **Scikit-learn**, **PyTorch**, or **TensorFlow** to train models.

Whatever option you go for, the purpose is that you are in full control over the training data, and you can choose which algorithms to use when training a model. This means it can take more skill, time, and compute power to get to the same end result as when we use out-of-the-box models. Both pre-built models as well as self-built models therefore serve their purpose and for each use case it should be evaluated which approach works best.

Summary

In this chapter, we discussed what skills you bring in as a data analyst, what you need to know about AI, and why combining AI and Power BI helps in the adoption of AI, and gave you a sneak peek into what we will cover in the rest of this book, namely the options for AI in Power BI. Not all options may be relevant for you or your organization, but at least subsequent chapters will provide you with a clear overview of what is possible so that you can educate yourself and others. Hopefully, this will give you the understanding of AI needed to help your company create a clear strategy on how to become a more data-driven organization that uses AI at a large scale.

In the next chapter, we will focus on the first input we need to train a model: the data. We will talk about how we can explore our data so that we understand what the input will be or what we need to do to make sure it is the right dataset for the job.

2

Exploring Data in Power BI

Once you get your hands on the data, it is time to do an initial exploration, also referred to as **exploratory data analysis (EDA)**. The goal is not only to get to know your data from top to bottom, to find outliers and missing data to fix during our data preparation, and to find correlations between features, but also to simply understand what the data looks like. By knowing your data, the model you will train is more likely to be accurate.

In this chapter, we will learn how we can use different techniques to explore our data before we use any of the **artificial intelligence (AI)** features in Power BI. By using the tools in Power BI that are covered in this chapter, we know what to do when preparing our data, and by understanding and cleaning up your data, you will ensure that you have good input for your model. Therefore, we will cover the following topics in this chapter:

- What to look for in your data
- Using data profiling tools
- Using visuals to explore your data

It is important to note that for large datasets, it is better to do any data exploration and preparation on the data source using `Python`, `R`, or **extract, transform, load** (ETL) professional tools. Exploring data in Power BI is a nice-to-have for easy and valuable insights any Power BI user can work with. Because of its simplicity and accessibility, we'll focus on the Power BI options. If you expect to work with large datasets or if you expect to do heavy computations on the data, it might be better to look for alternative tools before bringing the data into Power BI.

Technical requirements

There are three things you need to walk through the examples provided in this chapter, as follows:

- **Power BI Desktop**

  As this book revolves around Power BI, we expect you to have Power BI Desktop installed on your system. You can install Power BI from the Microsoft Store or find more advanced downloading options here: `https://www.microsoft.com/en-us/download/details.aspx?id=58494`.

- **Python**

  We will be using a Python visual in this chapter. To make sure you can view this visual, you need to install Python on your system, which you can download here: `https://www.python.org/`. Make sure to also install the `pandas` and `matplotlib` Python packages. You can do so by using `pip` in a console or shell, as follows:

  ```
  pip install pandas
  pip install matplotlib
  ```

- After you have installed Python and the two packages, you need to enable Python scripting in Power BI. You can do so by going into Power BI Desktop and selecting **File | Options and settings | Options | Python scripting**, as shown in the following screenshot:

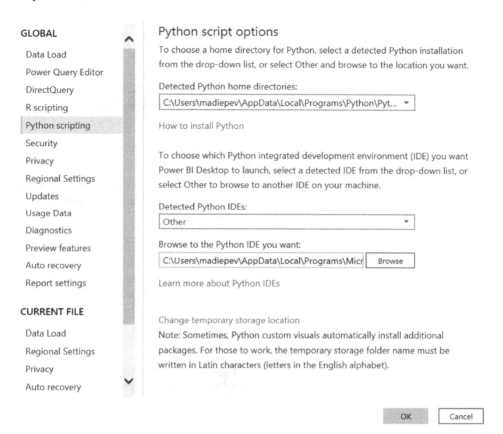

Figure 2.1 – Python script options

Find more detailed instructions on how to set up Python in Power BI in the online documentation at `https://docs.microsoft.com/en-us/power-bi/connect-data/desktop-python-scripts`.

- **Data**

  A sample dataset is used in the examples provided in this chapter. More details on the dataset are provided in the next section. You can download the dataset from GitHub. Go to the following link, right-click, and select the **Save as** option to save the data to your local drive: `https://raw.githubusercontent.com/PacktPublishing/Artificial-Intelligence-with-Power-BI/main/Chapter02/world-happiness-report.csv`

Using the sample dataset on world happiness

As explained in *Chapter 1*, *Introducing AI in Power BI*, we start our data science process by defining our use case. In this chapter, we will be using an existing use case with publicly accessible data so that we can reproduce the results. Every year since 2011, the *World Happiness Report* has used survey data from all countries in the world to determine where people are the happiest. You may have even seen these rankings being referred to in the news: `https://www.bbc.com/news/world-europe-56457295`. We will not get into whether this is a true measure of happiness; instead, let's focus on the data we have and explore that data.

Over the years, the *World Happiness Report* has used different ways of acquiring data. It has now decided to focus primarily on *Gallup World Poll* surveys. The original dataset can be downloaded from here: `https://worldhappiness.report/ed/2020/`. For any questions about collection or interpretations of the data, you can find **frequently asked questions (FAQs)** here: `https://worldhappiness.report/faq/`.

We will use a simplified version of this dataset that is provided in the GitHub repository for this chapter: `https://github.com/PacktPublishing/Artificial-Intelligence-with-Power-BI/blob/main/Chapter02/world-happiness-report.csv`. Compared to the original dataset, this simplified version contains fewer columns and, therefore, less data that we need to import into Power BI.

> **Important Note**
> This simplified version is a copy of the *World Happiness Report* posted on Kaggle here: `https://www.kaggle.com/ajaypalsinghlo/world-happiness-report-2021`.

How to interpret this dataset

Before we go on to explore the content of the dataset we will work with, make sure you download it first. You can find the dataset here: `https://github.com/PacktPublishing/Artificial-Intelligence-with-Power-BI/blob/main/Chapter02/world-happiness-report.csv`. Download it through this link, save it on your local system, and make sure to name the file `world-happiness-report.csv` as that is how we will refer to this local file from now on. You can open the **comma-separated values (CSV)** file to explore it and refer to it as we will first describe its contents.

The data included in this file is data used by the *World Happiness Report*. The first report appeared in 2011 and has been used every year to assess and compare the national happiness of countries. Since the data has been collected in a myriad of ways throughout the years, we will be using only the data collected from 2018 until 2020 as this represents the most consistent data.

The dataset we will use is based on *Gallup World Poll* surveys in which people from all countries are asked about how they evaluate their life. *They are asked to rate their life on a scale from 0-10, with 0 being the worst possible life, and 10 being the best possible life* (Source: `https://worldhappiness.report/ed/2020/`). The average score per country can be found in the `Life Ladder` column in our dataset.

To understand the `Life Ladder` score of each country, six factors are included in the data that may help explain or understand the score. These are outlined here:

- **Levels of gross domestic product (GDP)**: A measure of how much a country produces, which is often used as an indicator of a country's economic health

- **Social support**: National average of responses on survey questions related to whether respondents felt they can get support from their social circles, such as family and friends

- **Healthy life expectancy at birth**: Based on data from the **World Health Organization (WHO)**

- **Freedom to make life choices**: National average of responses to survey questions related to whether people felt they had the freedom to make their own life choices

- **Generosity**: National average of responses to survey questions asking whether participants have donated money to charity recently

- **Perception of corruption**: National average of responses to survey questions related to corruption throughout government and within businesses

And finally, the last two variables included in the data are two columns named `Positive effect` and `Negative effect`. These values are constructed from yes/no questions that ask respondents about their emotional experiences on the previous day. Scores are created by averaging answers across questions included in the survey and range from 0 to 1.

Importing the world happiness dataset into Power BI

You should have the `world-happiness-report.csv` file saved on your local system. We will import this file into Power BI to work with the data. We will now walk through importing the dataset and duplicating the data so that we have one query containing all years of data and one query containing only 1 year of data. This is to make subsequent examples easier to follow, without having to apply a filter every time we only want to visualize 1 year of data.

In the following steps, you will import the CSV file into Power BI and create two queries referring to it:

1. Select **Get Data**.
2. Select **Text/CSV**.
3. Navigate to where you stored the dataset. Select that dataset to open the file in Power BI.
4. A preview of the data should appear. Select **Transform Data**, which will open the **Power Query** editor.
5. Duplicate the table by right-clicking the table.
6. Rename the new table `world-happiness-2019`.
7. Apply a **filter** on the `year` column to only show rows for `2019`.
8. Select **Close & Apply** to load the data into your Power BI Desktop.

You have now imported the *World Happiness* dataset into Power BI. You should have two tables that are both queries on the same CSV file. For each country, we have data on the `Life Ladder` table. As mentioned earlier, the `Life Ladder` table consists of `Levels of GDP`, `Freedom to make life choices`, `Perception of corruption`, `Generosity`, `Social support`, `Healthy life expectancy at birth`, `Positive effect`, and `Negative effect` for the years 2018 to 2020. The second table contains the same data but is filtered for the year 2019.

We will be using this dataset in the examples to illustrate the concepts that will be explained in the rest of this chapter. We will explore this dataset to learn how to interpret its profile and learn how different visuals can give us more information about the data we have.

What to look for in your data

The foundation of AI is your data, which is exactly why we need to take very good care of our data. There are two main aspects that we need to investigate: the quantity and the quality of your data. One of the reasons AI is becoming an increasingly popular field is that its models are becoming better and easier to produce. This is partly because of how easy it is to get access to large amounts of data and process those large amounts of data to get a good model, thanks to cloud computing. *Garbage in is garbage out*, as they say, and the quality of your data is, therefore, just as important. We'll first talk about what we should do around data quantity, and then we'll discuss how we can explore and improve the data quantity.

Understanding data quantity

The purpose of a model is to *find a pattern in your data that can be generalized* to make interpretations about other or new data points. To make sure the model represents the true data well, we are inclined to collect as much data as possible. In many tasks that AI can do, this does indeed seem the case. AI can outperform humans very easily in repetitive tasks.

For example, the biggest predictor of someone installing solar panels on their roof is whether the neighbors have installed solar panels. Imagine a company selling solar panels that wants to target exactly those people with an offer for a solar panels installation. In many countries, it isn't required for people to report they have solar panels, so another way to check where there are solar panels is by using satellite images. The company could ask one of its employees to check all images for solar panels on roofs. Many of the images will be very similar, and the task of identifying solar panels is highly repetitive. This is a perfect example of a task where AI can quickly outperform humans, simply because a system using AI will be able to look at more examples in a shorter time span, to better learn how to differentiate between houses with or without solar panels.

So, what is the minimum amount of data you should have to train a model? This question is answered with the most unsatisfying of responses: *it depends*. The data we put in to train a model is also referred to as the training data and should be a good representation of the real data. If we want to train a model to detect tumors in X-ray images, we ideally want to have a model that is always correct, no matter what the tumor looks like. We can't collect images on all possible shapes and sizes that tumors can have, but we also don't want to train a model that only recognizes tumors if they are of a specific shape or size. Somewhere in between, there is this sweet spot of having data that reflects the many different types of tumors we may find without having to collect an endless amount of data. This trade-off depends on what it is you are training your model for. It also depends on whether you think your data indeed represents the real world and how much time and energy you want to spend on collecting the data.

In general, it is easier for models to work with large amounts of data to find patterns that we can use to generalize to a real-world application. However, we should stop worrying about getting more data as soon as we realize the model will not improve as a result and that it's just costing us time and money. One way to know if the model will indeed improve as a result of more data is by checking the data quality. We'll talk about how to explore that in the following sections.

Understanding data quality

The content of your data is maybe even more important than simply having a lot of data. There are many things we want to check for in our data to ensure it is good enough to use for AI purposes. The very first thing to do is understand your data and whether you have enough data to represent the real-world scenario you want to model. We will talk about how to evaluate whether you have a representative dataset through summary statistics and visualizations in Power BI in the following sections.

Representative dataset

In the previous section, we explored the importance of a **representative dataset**. Increasing the amount of data you use to train a model will only improve the model if the training data is a good reflection of the real situation. To know whether this is the case, we need to explore the training data to describe and understand it. This way, we can make better conclusions as to whether we indeed have a representative dataset.

To describe the data we work with, we can use **summary statistics**. We can use some very simple statistics to summarize the data we have and communicate insights from the data in the simplest way possible. The insights we typically get are outlined here:

- **Min**: The minimum or lowest value.
- **Max**: The maximum or highest value.
- **Mean**: The average value of all values in a column.
- **Median**: The middle number—the value at which 50% of the data is smaller and 50% of the data is larger.
- **Count**: The number of rows.
- **Empty count**: The number of rows that are empty and represent missing data.
- **Error count**: The number of rows that have data that is resulting in an error. For example, an error can mean you have specified the wrong data type on the column level. If you have text values in your column but you specified numbers, you will see ERROR in all rows where this is incongruent.

- **Distinct**: The number of different values in a column.

- **Unique**: The number of different values that only appear once in a column.

Generating the preceding insights of a dataset can give you a better idea of what the dataset contains. So, instead of pasting the whole table of our *World Happiness* dataset here, which would not even fit, let's have a look at summary statistics for the `Life Ladder` and the `Healthy life expectancy at birth` columns, which are shown in the following screenshot:

Statistic	Life Ladder	Healthy life expectancy at birth
Min	2.375	48.2
Max	7.889	77.1
Mean	5.612	65.359
Median	5.595	66.600
Count	381	381
Empty count	0	0
Error count	0	21
Distinct	361	179
Unique	343	78

Figure 2.2 – Table of summary statistics for Life Ladder and Healthy life expectancy at birth

From the preceding screenshot, we can see that the `Life Ladder` column contains values that range from `2.375` to `7.89`, even though the scale was 0-10. This, of course, makes sense, as people rarely will rank their happiness as an absolute 0 or 10. We tend to go for more nuanced answers. For `Healthy life expectancy at birth`, we see a minimum value of `48.2` and a maximum value of `77.1`, so we can safely assume this number represents age. With a mean of `65.359`, it does seem that there are more countries that have a higher age as a healthy life expectancy. To actually get more insights into the distribution of the data, we can visualize the data, as we will see in later sections of this chapter.

Figure 2.2 also shows that we count `381` rows for each variable. Therefore, we have 381 rows in our table, so it's not too big. Unfortunately, we also see we have `21` missing values in the `Healthy life expectancy at birth` column. At the same time, there are no missing values in the `Life Ladder` column. We will talk about missing data and what to do with it in the next chapter.

Finally, we can see the amount of distinct and unique values. For both these columns, it makes sense that both these values are pretty high as we have large ranges with small intervals, especially since we are using decimals. So far, nothing to worry about. If, for example, we had the same amount of unique values as we had values (which, in this case, means 381 unique values), then we would be dealing with what we call a high-cardinality feature. Models tend to not like having high-cardinality features as there is often no pattern to find in a feature that only has unique values—for example, a customer account number or an email address is a high-cardinality feature that probably won't provide any useful information for the model. You will likely want to remove high-cardinality features from your dataset as part of your data preparation.

Taking the *World Happiness* dataset as an example, we have now seen the first and easiest step you can take to explore your data. By looking at the summary statistics, you can get a quick glance at which columns your dataset contains and what are the contents of those columns. It can also give you an idea of possible problems that may exist in your dataset, which will show you what you have to do when preparing your data.

Using data profiling tools

When creating Power BI reports, we can visualize **key performance indicators** (**KPIs**) and important charts to help us understand the data we work with. Once we are working with the data as fields from our report view, we need to go back into the **Power Query Editor** to make any changes to our data. To avoid going back and forth between these two views, you already want to understand some basic things about your data while you are still in the **Power Query Editor** in order to clean up your data accordingly. To already explore your data in the **Power Query Editor**, we can use **data profiling tools**. These easy-to-use tools will show us what our data looks like so that we can decide what transformations we still need to apply in the **Power Query Editor**.

The data profiling tools consist of three parts, as outlined here:

- **Column quality**
- **Column distribution**
- **Column profile**

You can find these three features in the **View** tab of the **Power Query Editor**. There, you will find them in the **Data Preview** box, as seen in the following screenshot:

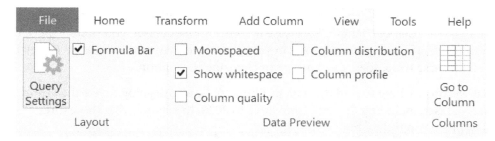

Figure 2.3 – View tab in the Power Query Editor showing the data profiling tools

You can see in *Figure 2.3* that you can check the box for which of the three data profiling tools you want to enable. You can choose whether you want to enable one, two, or three at the same time.

By default, Power BI will try to save on compute and only take the first 1,000 rows to generate this data preview. You can see this in the bottom bar of the **Power Query Editor**. If you want to run data profiling on all of your data, you can change this by selecting the **Column profiling based on 1000 rows** option. As you can see in the following screenshot, you can then change this to **Column profiling based on entire data set** by selecting that option:

Figure 2.4 – Options to base column profiling on 1000 rows or the entire dataset

In *Figure 2.4*, you can see the two aforementioned options, with the first being currently selected. Note that before changing it to base it on the entire dataset, you should check the size of your table. If you have many rows in your dataset, it may take a while for Power BI to process all these rows, so be cautious when choosing this option.

Now that we know where to find the data profiling tools, we are going to use the *World Happiness* dataset, including the data from 2018 to 2020, to explore what these three options to preview our data can do for us.

Column quality

The first and quickest check we can do on our data is to verify whether we have data in our table. Of course, a quick glance at your table can also help with this, but only for the first so many rows. To know whether all of your rows contain data for each column, you can check the box for **Column quality**, as you can see in the following screenshot:

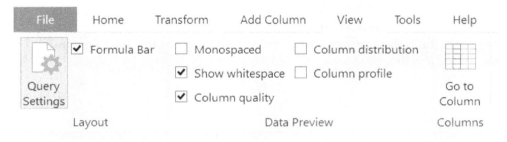

Figure 2.5 – Column quality selected

By checking the box, as you can see in *Figure 2.5*, you have **Column quality** selected. Once enabled, a new row will appear under your column names, as seen in the following screenshot:

⊞ ▾	Aᴮ꜀ Country name		¹²₃ year		1.2 Life Ladder		1.2 Log GDP per capita		1.2 Social support	
	● Valid	100%	● Valid	100%	● Valid	100%	● Valid	95%	● Valid	100%
	● Error	0%	● Error	0%	● Error	0%	● Error	0%	● Error	0%
	● Empty	0%	● Empty	0%	● Empty	0%	● Empty	5%	● Empty	0%

Figure 2.6 – Column quality results

Figure 2.6 shows the three types of rows you may have, as follows:

- `Valid`: The percentage of rows that contain valid data, ready to use for reports.

- `Error`: The percentage of rows that contain an error. Often, this means that there is data but it has not been loaded in properly, or the settings for the column have to be changed—for example, the date type set on the column level does not match the actual data.

- Empty: The percentage of rows that contain no data at all. This could indicate something went wrong with loading the data into Power BI, or it could be data that simply does not exist.

For the *World Happiness* dataset, we see there are some empty rows in the Log GDP per capita column. This means we have missing data for some reason. The other columns Country name, year, and Life Ladder seem to be fine so far, with 100% valid data.

Column quality gives us a quick preview of our data to check for any missing data or errors. Once we have inspected this for all our columns, we may want to explore our data even further and look at the **Column distribution** and **Column profile** options, as we will see in the following sections.

Column distribution

Next, to check for missing data in your table, it is also interesting to understand how many different values you have in your columns. For that, we can look at two metrics: **distinct and unique values**. If your column contains numbers, you may expect many distinct values; if you have text (such as Product Categories) and use that to categorize your data, then you may expect very few distinct values in your column. Next to that, if we want to use data for AI, we want to avoid high-cardinality features with many unique values, as explained earlier. Using **Column distribution**, we can check for the amount of distinct and unique values.

In the **View** tab in the **Power Query Editor**, you can check the box for **Column distribution**, as shown in the following screenshot, to view the amount of distinct and unique values:

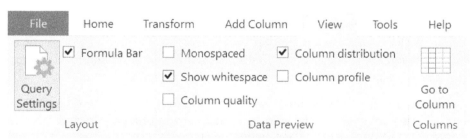

Figure 2.7 – Column distribution selected

In *Figure 2.7*, the checked box for **Column distribution** means that this is now enabled. As a result, we see a new window below each column, showing a histogram and the amount of distinct and unique values for each column, as shown in the following screenshot:

Figure 2.8 – Column distribution results

As you can see in *Figure 2.8*, the result is the amount of distinct and unique values, as well as a histogram showing the number of rows that contain a certain value. For example, for year, we see there are 3 distinct values and 0 unique values. This is expected, as we have the years 2018, 2019, and 2020, and we expected results for all countries for each of these years. However, looking at the histogram that consists of three bars above the distinct and unique count, we can see that around the same number of rows contain 2018 and 2019, but we have fewer rows containing 2020. This is worth further exploration as to why we have less data on 2020.

If we look at the Country name column, we can see we have 149 distinct and 7 unique values. So, there seem to be 149 different countries for which we collected data. However, there seem to be 7 countries that only appear once in this column, meaning we only have data for one of the 3 years for these countries, which could be related to our missing data in 2020. Again, this is something we could explore further.

For Life Ladder and Log GDP per capita, the distribution may not be as meaningful as we are working with numerical data here that is not categorical. Later in this chapter, we'll explore another approach to inspect the distribution of numerical variables, using histograms. For now, we do not see anything striking here, and having a lot of distinct and unique values is not a cause of concern for these columns.

We have seen how column distribution can give us more insights into the values we have in our columns. Interpreting the amount of distinct and unique values we have per column could show us potential problems with our data.

Column profile

The third and most extensive data profiling option is the **Column profile** option. This feature can give us the most insights into our data by calculating summary statistics and showing the **value distribution**.

Similar to **Column distribution**, the **Column profile** option also shows distinct and unique values but will give more detailed information when you select a column header.

In the **View** tab of the **Power Query Editor**, we can select **Column profile**, as shown in the following screenshot:

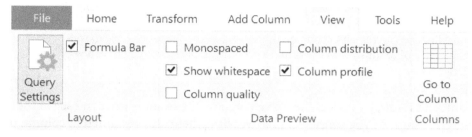

Figure 2.9 – Column profile selected

After **Column profile** is enabled by checking the box, we can then view the result for each column by first selecting one of the columns in the dataset. After column selection, an overview will appear at the bottom of the table, as shown in the following screenshot:

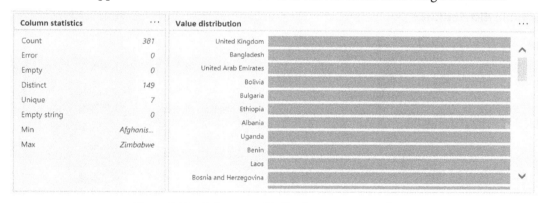

Figure 2.10 – Column profile for Country name column

On the left side of *Figure 2.10*, we can see **Column statistics**, which are the summary statistics. In other words, this is an overview of simple metrics to easily understand the contents of your data. We can see some values we already saw in **Column profile** and **Column distribution**, such as the amount of distinct and unique values. Some new insights are seeing the number of rows we have, as well as the average, minimum, and maximum value. Which statistics are shown and which should be explored depends on the type of data you have in your column.

> **The Default Column Profiling can be Misleading!**
>
> Remember the default for Power BI is to process the column profiling based on the top 1000 rows, which means you will see the count to be for 1000 rows. This, of course, is not the actual amount of rows in your dataset, but just the amount of rows Power BI calculated when processing the column profiling. If you want to know how many rows your dataset actually contains, change it to make sure Power BI takes the entire dataset when calculating the column profiling. Be careful with changing this setting as you may then have to wait for Power BI to finish with the column profiling, especially if you have a very large dataset.

To see a different result, let's select the `year` column. The result is what you see in the following screenshot. As you can see in this overview, there are now different column or summary statistics. The column year is of `date` type `Whole number` instead of `Text`, which means we have some other metrics we can look at, such as how many rows are `NaN` (which stands for **Not a Number**). Again, this could give us more insight into potentially missing or incomplete data. Following up on what we saw before with the **column distribution**, we again see that we do not have the same amount of rows for each year in the **value distribution**:

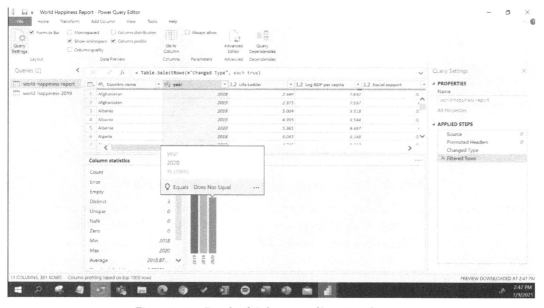

Figure 2.11 – Result of Column profile option for year

If you hover over one of the bars in the **Value distribution** view, you can see how many rows we have per year, as you can see for the year 2020 in *Figure 2.11*. We have 142 rows for 2018, 144 rows for 2019, and 95 for 2020. We do not exactly know why 2020 has fewer rows than the other years, but if we decided to exclude this year from our data, we could do so by right-clicking on this bar and choosing to apply a filter to not include this data in our table. In this way, we can very easily clean up our data based on the insights generated with the **Column profile** option.

We have now seen how **Column quality**, **Column distribution**, and **Column profile** are part of the data profiling tools in the **Power Query Editor** to give us quick insights into our data. By checking for missing data and understanding the values in our data and the shape of our data by looking at the distribution, we have taken the first steps to explore the data and are better able to decide how to clean it up and use it in our Power BI reports. We will talk more about the many things we can do to clean up our data in *Chapter 3, Data Preparation*. First, let's look into another way to get to know our data by using visuals to explore it.

Using visuals to explore your data

As a data analyst, you should be familiar with visualizing data. You use Power BI to make sure insights from the data are easily conveyed to your audience through visuals. The same visuals can also help you to understand the content and shape of your data.

First, we'll focus on the standard visualizations you can use in Power BI to explore your data, then we'll see how we can use Python in combination with the `matplotlib` library to create more customized visuals that we can add to our reports.

Line charts

One of the most basic charts we can create is a **line chart**. This is commonly used for showing how data changes over time to get insights into different trends.

To create a line chart, start in the **Report** view. We want to compare the `Life Ladder` score across the 3 years, so proceed as follows:

1. Select **Line chart** from the **Visualizations** pane.
2. Expand the `world-happiness-report` query in the **Fields** pane.
3. Drag `year` to the **Axis** field.

4. Drag `Life Ladder` to the **Values** field.

    We now see the sum of `Life Ladder` for 2018, 2019, and 2020 in the line chart. Instead of visualizing the sum of the `Life Ladder` score, it makes more sense to look at the average `Life Ladder` score. So, let's change this.

5. Select the down-pointing arrow next to `Life Ladder` in the **Values** field in the **Visualizations** pane.

6. Select **Average**.

The line chart now shows the average `Life Ladder` score per year for all countries, as illustrated in the following screenshot:

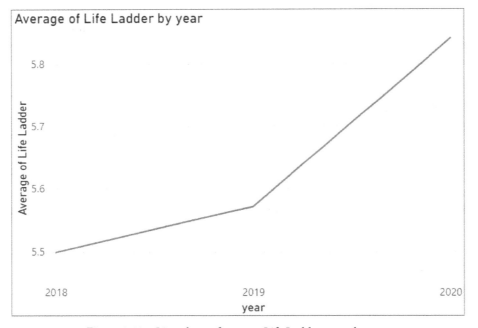

Figure 2.12 – Line chart of average Life Ladder score by year

As seen in *Figure 2.12*, the average score has increased over the years, which means that people across all countries were happier in 2020 compared to 2019 and 2018.

How you want to show your data insights and use them to tell a story is completely up to you. However, in data science, it is preferred to start the axis at 0 whenever it makes sense. As you may remember, the Life Ladder score could range from 0 to 10, so we can adjust the *y* axis to reflect this to see a better representation of the data, as follows:

1. Select the **Format** tab in the **Visualizations** pane.
2. Expand the **Y axis** tab.
3. Set **Start** to 0 instead of Auto.
4. Set **End** to 10 instead of Auto.

Now, we see that the average Life Ladder score is indeed still increasing over the years. The amount of increase, however, seems much less drastic in the following screenshot compared to what we saw in *Figure 2.12*. This could give us a more realistic view of the trends we visualize:

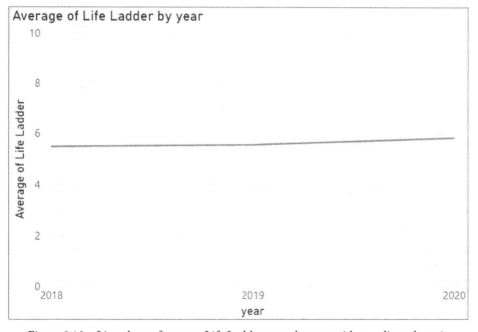

Figure 2.13 – Line chart of average Life Ladder score by year with an adjusted y axis

A line chart is a great way to visualize trends, most commonly used to show how a feature is changing over time or another feature with an order (such as days of the week). To get a representative view of the trends you visualize, make sure to adjust the *y* axis to at least start at 0 and optionally end at the maximum possible score.

Bar charts

When working with categorical features, it is convenient to use a bar chart. We can then visualize a quantity per category in a very intuitive way. Power BI offers two variations of a bar chart: a **bar chart** and a **column chart**. The bar chart shows the bar horizontally with the categories on the *y* axis, whereas the column chart shows the bar vertically with the categories on the *x* axis. They can be used for the same purpose, and it is up to you to decide which one you prefer for your report.

An example of what we could use a bar chart for is visualizing the `Life Ladder` score in 2019 per country included in the *World Happiness* dataset. We have many countries that potentially have long names, so we can save more space by creating a bar chart.

Start in the **Report** view, as follows:

1. Select **Stacked bar chart** from the **Visualizations** pane.
2. Expand the `world-happiness-2019` query in the **Fields** pane.
3. Drag `Country name` to the **Axis** field.
4. Drag `Life Ladder` to the **Values** field.

   We now see the count of `Life Ladder` for 2019 per country, which is equal to 1 for all of them. This does not give us much information. Instead, we want to look at the actual `Life Ladder` score per country for 2019, so let's change the value used for `Life Ladder`.

5. Select the down-pointing arrow next to `Life Ladder` in the **Values** field in the **Visualizations** pane.
6. Select **Average**.

As a result, we will see a bar chart, as shown in the following screenshot:

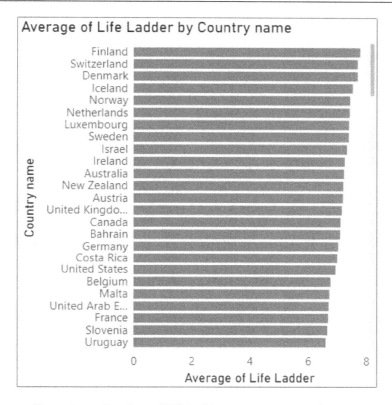

Figure 2.14 – Bar chart of Life Ladder score per country for 2019

Showing all countries probably does not fit in your screen, which is why we see a scroll bar on the right side of the visual with which we can scroll through all scores. In this case, the *x* axis already automatically starts at 0, so we do not have to change it. As a best practice, make sure you are aware of the range on your axes and change them to start at 0 to give you a clear view of the data you want to compare.

We have now seen how we can use a bar chart to compare values for different categories. In data science, we often use a special type of bar chart to explore the distribution of the data. This is known as a histogram, which we will explore in the next section.

Histograms

A **histogram** is actually a special type of bar chart. The special thing we do in histograms is that we group our data into buckets and then count how many values are included in each bucket. By plotting the number of values per bucket, we can see *how the data is distributed*, which is important for deciding whether you indeed have a representative dataset (as explained in the first section of this chapter).

A histogram is perfect for visualizing the distribution of numerical data, whereas a bar chart is ideal for categorical data. The first thing we have to do in Power BI to be able to create a histogram is to create buckets we want to visualize. We can do this by grouping the data into bins. The size of the bins is something you can experiment with and edit in Power BI after creating a histogram. Although there have been attempts at creating a golden rule for what the size should be, an easy way to decide on the bin size is to take the range of possible values and divide it by 10 to ensure you will see 10 bars in your histogram. If you then feel you are missing some information, you can decrease the bin size to increase the number of bars you will see in your histogram.

Let's see how we can create a histogram in Power BI, as follows:

1. In the **Fields** pane, expand the `world-happiness-2019` query.

2. Select the three dots (**…**) to the right of `Life Ladder` to display the **More options** menu.

3. A menu of options appears. Select **New group**, as shown in the following screenshot:

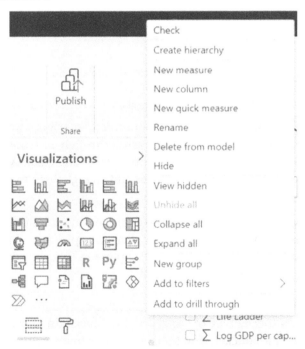

Figure 2.15 – More options menu of Life Ladder

Power BI will give you a suggestion for bins. We want to see the frequency of values for each bucket of the Life Ladder score. To make our histogram easy to interpret, let's create buckets to include all possible scores and give them a size value of 1. This means our first bucket will contain all scores from 0 to 1, the second bucket contains the count of scores from 1 to 2, and that is how we will continue onward.

4. Change the **bin size** value to 1.

5. Select **OK**.

    We created a new field in the world-happiness-2019 query called Life Ladder (bins). As mentioned before, it may be that you later want to change the size of the bins. You can still do so after you have created a histogram. Be aware that the bin size is defined for the field, so any histogram you have that uses the field will use the new bin size. If you want to create two histograms showing different groupings of the same data, then you have to create a new group, as described in the preceding steps. Alternatively, you can edit the Life Ladder (bins) field by selecting the three dots (…) next to this new field and choosing the **Edit groups** option.

    Back to the creation of our histogram. We will now use the Life Ladder (bins) field we just created instead of the original Life Ladder field to create a bar chart. Histograms normally show vertical bars, which is why we will use a column chart this time.

6. Select **Stacked column chart** from the **Visualizations** pane.

7. Drag Life Ladder (bins) to the **Axis** field.

8. Drag Country name to the **Values** field.

    Power BI automatically visualizes the *count* of Country name for each bin we created. This is indeed what we want to see. We want to see the number of rows within a certain bin, which also means we could use another feature instead of Country name too. As long as we see the count, the histogram should make sense.

    You want to also include the buckets for which you have a count of 0 to make sure all possible values are shown. We can visualize this by changing the *x* axis.

9. Select the **Format** tab in the **Visualizations** pane.

10. Expand the **X axis** tab.

11. Set **Start** to 0 instead of Auto.

12. Set **End** to 10 instead of Auto.

The result is a histogram, as seen in the following screenshot. We can now see the distribution of values for the Life Ladder score for 2019. Based on the shape of the histogram, you can see what type of distribution you are dealing with. If it is bell-shaped, like the one shown here, it is said to be approximately **normal**. Very often, this is what we like to see. It tells us that most of the Life Ladder scores are close to the average or mean, which is 5.57, and only a small subset of scores is at the lower end (toward 0) or higher end (toward 10) of the scale we used:

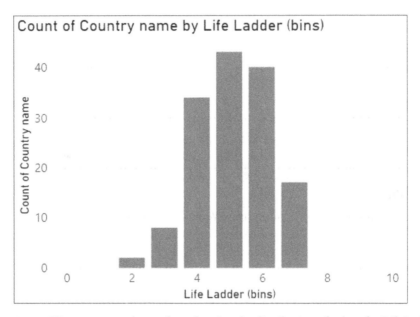

Figure 2.16 – Histogram or column chart showing the distribution of values for Life Ladder

Understanding the distribution of your data can tell you whether you indeed have a representative dataset. If, for example, we create a histogram of Healthy life expectancy at birth, as shown in the following screenshot, we can see a slightly different distribution. It is not a symmetric bell shape; instead, there would be more countries where the healthy life expectancy is at the higher end of the scale. This is called a **left-skewed distribution**. In this case, the average or mean (65 years) is smaller than the middle value or median (67 years). This could indeed be a good representation of reality, and as long as it makes sense for the data you are looking at, there should be no cause for concern:

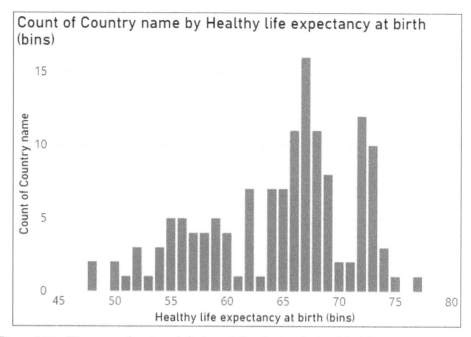

Figure 2.17 – Histogram showing a left-skewed distribution for Healthy life expectancy at birth

Finally, we can have a **right-skewed distribution** if we have a higher count of values at the lower end of the scale. As a result, the mean will be larger than the median. This is the case for the Negative effect column in our *World Happiness* dataset. With a mean of 0.24 and a median of 0.20, the mean is larger than the median.

Interestingly, if we bin the `Negative effect` column using a bin size of 0.1 and visualize the count of `Country name` per bin in a column chart, we may not immediately see that it is a skewed distribution. It is only when we adjust the *x* axis to start at 0 (the lowest possible score) and end at 1 (the highest possible score) that we see it is not normally distributed but right-skewed, as shown in the following screenshot. This again highlights the importance of choosing the right range for your axes:

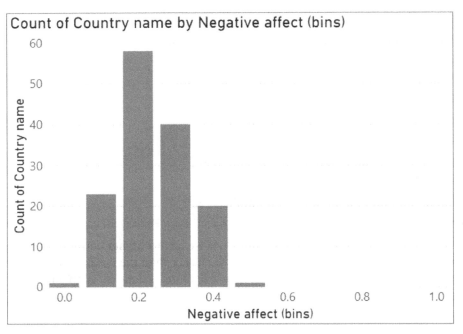

Figure 2.18 – Histogram showing a right-skewed distribution for Negative effect

We have seen how we can create a histogram in Power BI by grouping our numerical data and using a stacked column chart. This will help us to explore the distribution of our data to determine whether we indeed have a representative dataset. We will later see how we can also create a histogram using a Python visual and the `matplotlib` library in Power BI.

Scatter plots

So far, we have mostly been looking at features in isolation from each other. The reason we want to use AI on a dataset is that we want to find patterns in our data and learn from them in the future. That also means we should be interested in how features are related to each other. Some relationships may be very straightforward—for example, when the weather gets warmer, we expect to sell more ice creams. Even though this is an intuitive relationship, we want to actually have the data to prove it. **Scatter plots** can be used to plot two features or fields against each other to find out whether there is a correlation between the two.

To create a scatter plot, go through the following steps in Power BI:

1. Select **Scatter chart** from the **Visualizations** pane.

2. Expand the `world-happiness-2019` query to view its fields.

3. Drag `Life Ladder` to the **X Axis** field.

4. Drag `Healthy life expectancy at birth` to the **Y Axis** field.

> **Adjust the Axes to have a Realistic View of your Data**
>
> To get a realistic view of your data, always double-check the range on your axes to see if they make sense. Power BI by default chooses the smallest range to show all your data points, but this may result in a distorted view of the data. An increase in data may, for example, appear to be very large. In reality, it is only a small increase compared to the true scale. In this example, we keep the axes as-is since it will not make much of a change to the conclusions we will draw from the visual.

Optionally, we can add a trend line to see the relationship more easily between the two fields.

5. Go to the **Analytics** tab in the **Visualizations** pane.

6. Expand **Trend Line**.

7. Select **+ Add**.

As a result, you will see a scatter plot with a trend line, as shown in the following screenshot, going from the left bottom of your chart to the right top. This shows a **positive correlation** between `Life Ladder` and `Healthy life expectancy at birth`. In other words, when there is a higher life expectancy, there is a higher score on the `Life Ladder` axis. Be very careful with these kinds of conclusions as a correlation never proves cause and effect. We can always assume based on our domain knowledge, but a scatter plot does not prove which feature is the cause or effect:

Figure 2.19 – Scatter plot showing a positive relationship between Life Ladder and Healthy life expectancy at birth

An upward line shows a positive correlation between the two features. A downward line that goes from the left top of the chart to the right bottom indicates a **negative relationship**. We can see an example of this in the following screenshot. When there is more perception of corruption, we see a lower ranking in the Life Ladder axis. In this case, looking at the data points in the scatter plot may not have made this relationship very clear, which is why a trend line can help to see what the overall trend is in the relationship between these two fields:

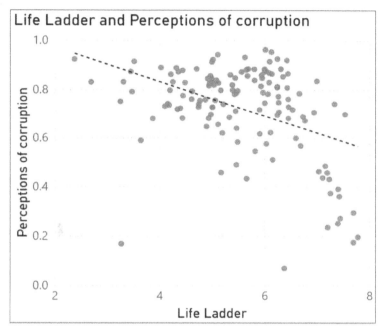

Figure 2.20 – Scatter plot showing a negative relationship between Life Ladder and Perceptions of corruption

Exploring data with scatter plots can be challenging. Sometimes, you may be tricked by the data, which is why you should always interpret these charts with the help of some domain knowledge. Take the scatter plot shown in the following screenshot, for example, where we look at the relationship between generosity and healthy life expectancy at birth. In this case, there is somewhat of a decreasing trend line. However, taking the randomness of the data points into consideration, the safest conclusion here is that there is **no relationship** between these two features. Simply looking at the trend line, possibly in combination with the wrong scale at one of the axes, could indicate a correlation. This is why you must think about what makes sense to explore and look carefully at the charts you create:

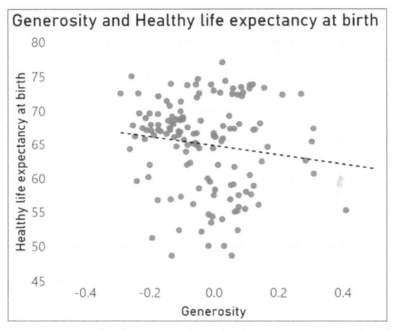

Figure 2.21 – Scatter plot showing no relationship between Generosity and Healthy life expectancy at birth

We have now seen how we can use scatter plots to explore whether there is a positive or negative relationship between two features. We have also learned again how important it is to adjust the scale of the axes before drawing any conclusions.

matplotlib

As we are dipping our toes in the field of AI, it can't hurt to learn a bit more about the preferred coding language of a data scientist: **Python**. Don't worry—you don't need any prior experience with Python to be able to plot some special charts in Power BI using a Python visual. However, the reason we still want to explore this option is that for some data exploration, we want to visualize the data in certain ways.

We already talked about histograms, which we can create by grouping data and using a stacked column chart. Using a Python visual, we can use the matplotlib library to easily create those buckets and histograms for us. Next to that, there is a **box plot**, which is not a standard visualization in Power BI, but again is easily created using a Python visual. Other plots such as line and bar charts can also be created with matplotlib, but since we can also easily do that with the standard visualizations of Power BI, we'll focus on histogram and box plots to see how we can use Python in Power BI. It is, of course, up to you whether you would want to explore this option further.

To walk through these examples yourself, make sure you have Python and the pandas and matplotlib libraries installed on your system and have connected Power BI to your installation. Find instructions on how to do this in the *Technical requirements* section at the beginning of this chapter.

Histogram

To visualize the distribution of your data, you can use a histogram. Earlier, we saw how we can use some of Power BI's features to create a histogram: grouping your data into bins and visualizing that data with a stacked column chart. Next to that, we can also create a histogram using a Python visual. We'll be recreating similar histograms as what we created before so that you can easily compare the two options. Follow these next steps:

1. From the **Visualizations** pane, select **Python visual**.

2. From the world-happiness-2019 query, drag Life Ladder to the **Values** field.

3. In the **Python script** editor, leave the commented code and add the following script:

```
import matplotlib.pyplot as plt
bins = [0,1,2,3,4,5,6,7,8,9,10]
plt.hist(dataset['Life Ladder'], bins, histtype='bar',
rwidth=0.8)
plt.xlabel('Life Ladder')
```

```
plt.ylabel('Number of countries')
```

```
plt.show()
```

In the preceding code snippet, we import the `matplotlib` package and refer to the `pyplot` functionality with `plt`. We then specify the bins and use `pyplot` (as `plt`) to plot a histogram with `Life Ladder` as the values, using the bins specified in the preceding line. There are some extra parameters you can change, such as the width of the bars (`0.8`), the x label (`Life Ladder`), and the y label (`Number of countries`).

Your script editor should look like this:

Python script editor

⚠ Duplicate rows will be removed from the data.

```
2
3  # dataset = pandas.DataFrame(Life Ladder)
4  # dataset = dataset.drop_duplicates()
5
6  # Paste or type your script code here:
7  import matplotlib.pyplot as plt
8  bins = [0,1,2,3,4,5,6,7,8,9,10]
9  plt.hist(dataset['Life Ladder'], bins, histtype='bar', rwidth=0.8)
10 plt.xlabel('Life Ladder')
11 plt.ylabel('Number of countries')
12
13 plt.show()
```

Figure 2.22 – Python script editor

After you have pasted the code into the **Python script editor**, the visual will likely automatically show the result. Just to be sure that you can force it to run the script.

4. Click on the **Run script** icon at the right top of the **Python script editor** to ensure the script is executed. A visual will then appear in your report, as shown in the following screenshot:

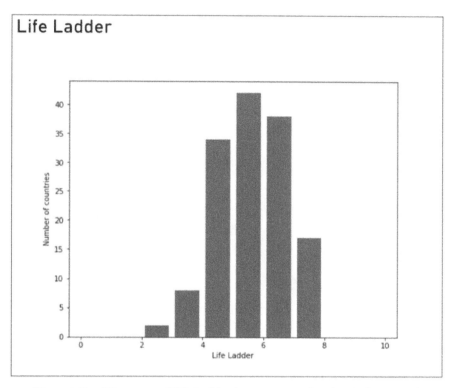

Figure 2.23 – Histogram of Life Ladder in 2019, created with a Python visual

Figure 2.23 shows a histogram created with a Python visual. The following screenshot shows the histogram we created earlier. Since we chose the same bin size, we should see the same distribution in both charts. There are a few design changes that come from one being created by a Python library and one being created by Power BI. However, the insights should be the same:

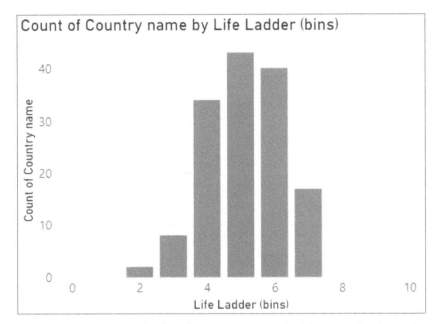

Figure 2.24 – Histogram of Life Ladder in 2019, created with a stacked column chart

As you may have noticed from the code, we defined our bin size by creating a `bins` variable. The content of this variable was a list where we indicate the range of data (0-10) and the interval (1). If we would want to decrease or increase the bin size, we could change the content of `bins` to see a histogram with more or less detail, respectively.

For example, we could only update *line 8* in the code and use this instead:

```
bins = [0,0.5,1,1.5,2,2.5,3,3.5,4,4.5,5,5.5,6,6.5,7,7.5,8,8.5,
9,9.5,10]
```

In that case, we would get a histogram with more detail, as shown in the following screenshot:

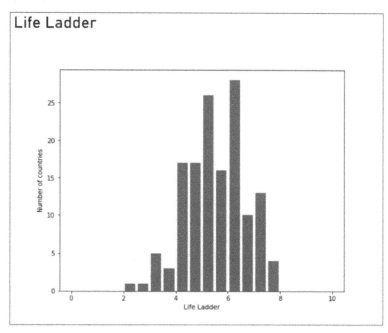

Figure 2.25 – Histogram with more detail of Life Ladder in 2019

There are many parameters we can adjust to see a slightly different histogram, such as the color and width of the bars. To view everything you can change in a histogram plot, see the `matplotlib` documentation at `https://matplotlib.org/3.1.1/api/_as_gen/matplotlib.pyplot.hist.html#matplotlib.pyplot.hist`.

By using a Python visual in combination with `matplotlib` in Power BI, we have another way to create histograms. Whether you prefer this method or using a stacked column chart instead is completely up to you.

Box plots

We have previously talked about the importance of looking at summary statistics and how these can tell us whether we have a representative dataset. The built-in data profiling tools we have in the **Power Query Editor** allow us to quickly view some of the summary statistics per column to explore our data. Another way to view summary statistics is to visualize them in a box plot.

A **box plot** is another chart used to visualize the shape or content of your numerical data. It does so by taking the following six metrics of the summary statistics: the minimum value, the lower quartile (value at 25%), the mean (average value), the median (value at 50%), the upper quartile (value at 75%), and the maximum value. By visualizing these metrics, we can easily see how the data is divided into four sections that each contain 25% of the data.

There is no Power BI visualization that allows us to create a box plot. There are some options that we can get from **AppSource**. However, those have low ratings at the time of writing and are not managed by Microsoft, which means not all organizations will allow the use of these visuals. Therefore, let's now look into how we can create a box plot with a Python visual, as follows:

1. From the **Visualizations** pane, select **Python visual**.

2. From the world-happiness-2019 query, drag Life Ladder to the **Values** field.

3. In the **Python script editor**, leave the commented code and add the following script:

```
import matplotlib
dataset.boxplot("Life Ladder", showmeans=True,
showfliers=False)
plt.show()
```

You should now see a box plot, as shown in the following screenshot. The horizontal lines at the bottom and top are the minimum and maximum values, respectively. These lines are often also referred to as *whiskers*. The box shows the middle 50% of the data, meaning that the bottom line is the lower quartile and the upper line is the upper quartile. The line in the middle is the median, and the triangle in the middle is the mean. We looked at the distribution of Life Ladder before and since the median and mean are very close together, and the box looks fairly symmetrical, so we can assume it is **normally distributed**:

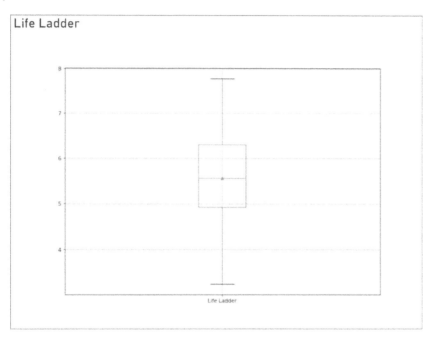

Figure 2.26 – Box plot of Life Ladder created with a Python visual

Let's look at another box plot, this time of `Healthy life expectancy at birth`, as shown in the following screenshot. First of all, we can see that the mean is smaller than the median, which indicates a **left-skewed distribution**. This is in line with what we concluded from the histogram we plotted in *Figure 2.17*. We can also see that the median is not nicely in the middle of the box, showing again that the data is not symmetrical:

Figure 2.27 – Box plot of Healthy life expectancy at birth created with a Python visual

Looking at the distribution of your data with the help of summary statistics can tell you a lot about the shape of your data. To visualize this information, we can create a box plot using the `matplotlib` library and a Python visual in Power BI.

When working in Power BI, we can use many of its standard visuals to explore our data. Another option is to use Python visuals, whereby we use the `matplotlib` Python library to visualize our data. The two most common visuals when exploring your data are a histogram and a box plot, both being great tools to understand the distribution of your data.

Summary

In this chapter, we learned how to quickly view important metrics to understand the contents of your data by generating summary statistics with data profiling tools (**Column quality, Column distribution**, and **Column profile**) in the **Power Query Editor**. We then discussed the many different visualizations that can be used to explore your data, such as line, bar, column, and scatter charts. We used a Python visual and learned how to create histograms and box plots with the `matplotlib` library. All these tools will help you to understand your data, to learn whether it is a representative dataset that you should continue to use, and as you get your first insights from your data, you will be able to judge how it can be used for different AI projects.

Now that we understand the content of our data, we know that there are some problems to fix before we move on to AI. To ensure data quality, we need to fix our outliers, missing data, and imbalanced data that can negatively influence the accuracy of any model we want to train. We will discuss how to do that in the next chapter.

3
Data Preparation

The most important input for **Artificial Intelligence** (**AI**) is the data you have. You must ensure that you have enough data and that it is of good quality. The first step we take after importing data into Power BI is exploring the data to make sure we understand what we are working with. The next step is to clean the data and prepare it for our AI applications.

In this chapter, we'll learn about the common problems we may have with our data and how to fix them. It is important to prepare the data properly in the **Power Query Editor** as this will help us in extracting the right insights from the data using AI. The common problems we will talk about in this chapter are as follows:

- Fixing the structure of your data
- Working with missing data
- Mitigating bias
- Handling outliers

Fixing the structure of your data

The data you connect to in Power BI will appear in *tabular format*. For example, when we import data from an Excel file into Power BI, we can view it as a table with **column headers** and **row values**. Based on the data detected and the information gathered from the data source, Power BI will then also assign a **data type** to each column. You can change the data type for each column afterward, and this is necessary for some features you may want to use. To calculate certain *summary statistics* such as averages, as we have seen in *Chapter 2, Exploring Data in Power BI*, we need a column to have numerical data as the data type. To do *time-series forecasting*, as we will see in *Chapter 4, Forecasting Time-Series Data*, we need a column that has date/time as a data type.

In the following figure, you can see the design of a table as presented in Power BI and the important features:

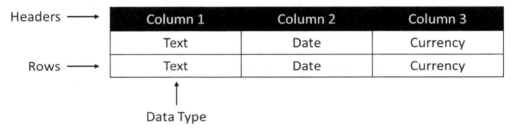

Figure 3.1 – Design of a table in Power BI

As shown in a simplified way in the preceding table, data will be shown in a tabular format in Power BI. This means we have columns, each having a header that will appear at the top of our table. The columns will contain values or rows, and each value within a column must be of the same data type. As shown in the figure, if you want to specify that **Column 2** contains date values, all rows much match the same formatting of this type of data. Otherwise, you will see an `Error` value.

Why is this so important to understand? Because, when we are working with AI, we may not always have data that intuitively would fit into a table. However, Power BI really wants us to fit the data we work with into a table. As a result, whether working with **structured**, **semi-structured**, or **unstructured** data, we need to present the data to Power BI in a table with column headers and rows that adhere to the data type we specify. In subsequent sections of this chapter, we'll cover these three types of data.

Structured data is already in a tabular format at the source. Think of data you have stored in a SQL database, in an Excel file, or even in a **Comma-Separated Values** (**CSV**) file. However we have stored the data, it is very easy to read this data as tabular data. Semi-structured data, on the other hand, is often based on keys and values. This data is not stored in a tabular format in the source, but rather as JSON files or documents. Storing data as semi-structured in, for example, an Azure Cosmos DB database gives us more flexibility with the data we want to store but leaves us with more work when we want to convert it into a table to use in Power BI. Lastly, we can have unstructured data, such as images or text, that we want to visualize in Power BI. We can use images in our reports to make them more visually appealing. But we can also use AI to extract tags from images (*Chapter 10, Getting Insights from Images with Computer Vision*) and keywords from text (*Chapter 8, Integrating Natural Language Understanding with Power BI*) to visualize trends we see in these kinds of unstructured data.

Before we can extract any insights from our data, whatever type of data we have, we need to import the data and prepare it according to its structure. The steps taken depend on the type of data you work with, which is why we will cover the three types, structured, semi-structured, and unstructured, one by one using common examples. We will first focus on how to fix the structure, and we will then continue with fixing problems we may have with the contents of the table, which can occur with any of these types.

Working with structured data

We will start with the type of data Power BI likes best: structured data. Whenever we talk about structured data, we mean data that often has a *predefined schema*. We know which columns we have, their names, and the type of data we have in each column. This makes it very easy to populate our table with values in the rows, as we know exactly which data goes where.

As long as Power BI can easily detect what the columns are, their names, the data type, and which values should be recognized as rows, your data will probably be imported correctly. However, there are some data sources that can be seen as structured, where we do need to make sure Power BI understands how to read the data. One such example is a CSV file, where we have to tell Power BI how to recognize individual columns, what the column names are, and potentially, fix the data type of the columns.

Let's take an example of a CSV file and look at the different things we may have to fix to ensure the structure is fixed. We will take a CSV file containing data on the *World Happiness Report* over recent years (for more information on the dataset, see *Chapter 2, Exploring Data in Power BI*). We will import it to see the different options for reading the CSV file, and most importantly, we will look at the options we have to *fix the structure* of the table in the Power Query Editor.

For this example, we'll use the same dataset as we were working with in *Chapter 2, Exploring Data in Power BI*. If you've already imported the data in Power BI, you may skip the first seven steps. If you haven't imported the data yet, download the dataset as a CSV file from `https://raw.githubusercontent.com/PacktPublishing/Artificial-Intelligence-with-Power-BI/main/Chapter03/world-happiness-data.csv`. Save the file locally on your system, name it `world-happiness-data.csv`, and make sure you can navigate easily to it in the next steps:

1. In **Power BI**, select **Transform Data**.
2. Of the options under **Transform Data**, select **Transform Data** to open **Power Query Editor**.
3. In **Power Query Editor**, select **New Source**.
4. Select **Text/CSV**.
5. Navigate to the CSV file you downloaded and named `world-happiness-data.csv`.
6. A popup should appear as you can see in the following figure:

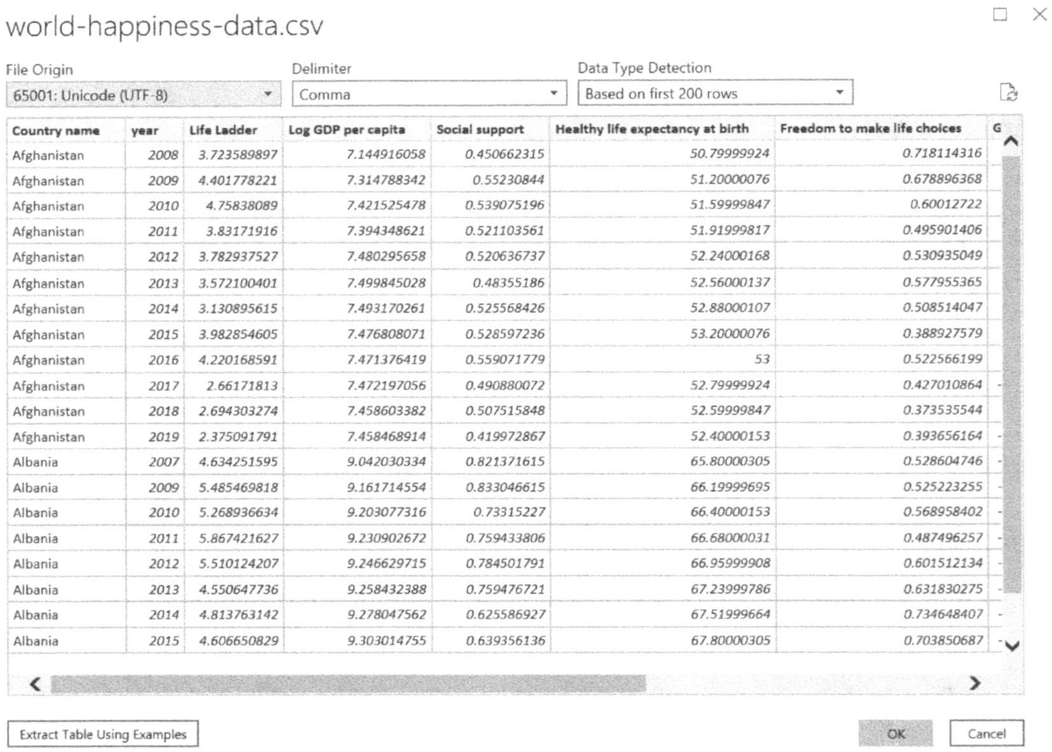

Country name	year	Life Ladder	Log GDP per capita	Social support	Healthy life expectancy at birth	Freedom to make life choices
Afghanistan	2008	3.723589897	7.144916058	0.450662315	50.79999924	0.718114316
Afghanistan	2009	4.401778221	7.314788342	0.55230844	51.20000076	0.678896368
Afghanistan	2010	4.75838089	7.421525478	0.539075196	51.59999847	0.60012722
Afghanistan	2011	3.83171916	7.394348621	0.521103561	51.91999817	0.495901406
Afghanistan	2012	3.782937527	7.480295658	0.520636737	52.24000168	0.530935049
Afghanistan	2013	3.572100401	7.499845028	0.48355186	52.56000137	0.577955365
Afghanistan	2014	3.130895615	7.493170261	0.525568426	52.88000107	0.508514047
Afghanistan	2015	3.982854605	7.476808071	0.528597236	53.20000076	0.388927579
Afghanistan	2016	4.220168591	7.471376419	0.559071779	53	0.522566199
Afghanistan	2017	2.66171813	7.472197056	0.490880072	52.79999924	0.427010864
Afghanistan	2018	2.694303274	7.458603382	0.507515848	52.59999847	0.373535544
Afghanistan	2019	2.375091791	7.458468914	0.419972867	52.40000153	0.393656164
Albania	2007	4.634251595	9.042030334	0.821371615	65.80000305	0.528604746
Albania	2009	5.485469818	9.161714554	0.833046615	66.19999695	0.525223255
Albania	2010	5.268936634	9.203077316	0.73315227	66.40000153	0.568958402
Albania	2011	5.867421627	9.230902672	0.759433806	66.68000031	0.487496257
Albania	2012	5.510124207	9.246629715	0.784501791	66.95999908	0.601512134
Albania	2013	4.550647736	9.258432388	0.759476721	67.23999786	0.631830275
Albania	2014	4.813763142	9.278047562	0.625586927	67.51999664	0.734648407
Albania	2015	4.606650829	9.303014755	0.639356136	67.80000305	0.703850687

Figure 3.2 – Import options for the CSV file

In this screenshot, you see the options presented to you in Power BI when importing data from a CSV file. The most important thing we need to specify is how the columns are delimited. In this case, **Delimiter** is **Comma** as is configured at the top of the pop-up screen shown. If you choose the wrong delimiter, you might not see multiple columns, but instead will only see one column. Change the type of delimiter to make sure you import the data correctly.

7. Select **OK**.

We have now imported the data from a CSV file. There are three things that are important when importing data as a table:

- *Having the right column headers*: If the headers are on the first row of the table instead of at the top of the table, you can fix this by selecting **Use First Row as Headers** in the **Home** ribbon. In general, a best practice in Power BI is to make sure the headers have names that are intuitive and easily readable. For example, use Country name instead of abbreviations such as CN or jointed words such as CountryName. Since the column headers are shown as the field names in your reports, meaning the consumers of the reports will see these names, you need to make sure you make it understandable for them.

- *Having the right columns*: We are using a CSV file. This means columns are delimited with a comma. We have specified this when importing the data to make sure Power BI reads the data correctly. If you are using a different delimiter, you can change this configuration when importing data. Alternatively, you can **split columns** in **Power Query Editor** based on different criteria such as delimiter, number of characters, or digit and non-digit.

- *Having the right data type specified per column*: In the following figure, we see the data we have imported from the CSV file containing the *World Happiness Report* data:

	ABC Country name	123 year	1.2 Life Ladder
1	Afghanistan	2008	3.723589897
2	Afghanistan	2009	4.401778221
3	Afghanistan	2010	4.75838089
4	Afghanistan	2011	3.83171916
5	Afghanistan	2012	3.782937527
6	Afghanistan	2013	3.572100401
7	Afghanistan	2014	3.130895615
8	Afghanistan	2015	3.982854605
9	Afghanistan	2016	4.220168591

Figure 3.3 – Imported data from CSV file

Here, we see the `Country name`, `year`, and `Life Ladder` columns. Next to `Country name` we see **ABC**, indicating that the data type in this column is text. For `year` we see **123**, which means these are whole numbers, and for `Life Ladder` we see **1.2**, which tells us the data type of this column is a decimal number. In *Figure 3.4*, we see what happens if we click on this icon showing either **ABC**, **123**, or **1.2** (or any of the other options depending on the recognized data type):

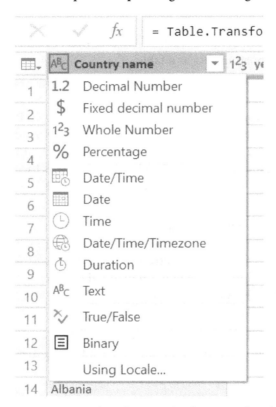

Figure 3.4 – Options when changing the data type of a column

For each column, you can always change the data type in the Power Query Editor, by clicking on the icon to the left of the column name as shown here. You can then choose any of the options shown in the same figure, and even choose the locale to make sure you use the right formatting. You can change the data type at any point later if you notice visuals or features do not work. The most important thing is to make sure the data is correctly identified as either of numbers, date/time, text, or binary (0 or 1, or true or false).

We have seen how to import structured data by taking an example of a CSV file. We have discussed the three most important things to validate after data is imported:

- Whether the column headers are correct
- Whether you have the right columns
- Whether you have the right data type configured on the column level

Next, we will see how we can import semi-structured data and work with it in a tabular format in Power BI.

Fixing the structure of semi-structured data

Even if our source data is not structured, we can still work with it in Power BI. Before we look into how we can fix the structure of semi-structured data from an Azure Cosmos DB database in the Power Query Editor, let's talk a little bit more about what this type of data can look like.

The reason we want to store semi-structured data is often that we want to be flexible with the variables we collect. Take a webshop as an example. You sell articles of clothing that have a lot of information you want to store. For pants, you need to know the size, the color, and the brand, while for shirts, you may want to know these things too, as well as the type of neckline, for example, a V-neck or a round neck. If we stored this data in a database where we had to define a schema before storing the data, we would have to go back and change the schema every time we had a new variable we wanted to store.

What if we are now also showing for each article of clothing whether it is sustainable or not? We may not have this information for the clothes we were selling so far, but we want to add it as another tag to the items we are adding to our webshop. To make sure we keep this flexibility of deciding what to store when we store it, we use semi-structured data. This is data that you can still query, but it is stored as **key-value pairs**, such as `Product Category` as the key and `pants` as the value, instead of as a table.

Whenever we are using this type of data in Power BI, we have to specify which keys we want to include in our table. The key will become the column header, and Power BI will populate each column with all the values it can find for that key to create the table. This may result in an incomplete table, so this has to be done deliberately and carefully.

Let's look at an example of importing and preparing semi-structured data in Power BI to see what things we may need to fix. For this example, we will use data collected from Twitter. A dataset has been created by storing all tweets that contain the words *artificial intelligence* in the period of July 14 2021 until July 31 2021.

Using the Twitter API, you can retrieve tweets based on certain keywords or hashtags. The Twitter API returns semi-structured data. In this example, the data is saved as a local JSON file.

The data used in the example is available to download here: `https://raw.`
`githubusercontent.com/PacktPublishing/Artificial-Intelligence-`
`with-Power-BI/main/Chapter03/aitweets.json`. Save the file as `aitweets.`
`json` on your local system:

1. In **Power Query Editor**, select **New Source**.

2. Select **More…**.

3. Select **JSON**, and select **Connect**.

4. Navigate to the **aitweets.json** file you downloaded and saved locally. Select this file for Power BI to import.

5. If the **Connection settings** window appears, select **Import**. Then, select **OK**.

In the next screenshot, we see the result after importing the data from the local JSON file:

$^{A^B}_C$ AuthorLocation	CreatedAt	$^1{}^2{}_3$ RetweetCount	$^{A^B}_C$ TweetLanguage	$^{A^B}_C$ TweetText
	7/14/2021 2:47:39 PM	0	en	@____bruvteresa_ Accordi
Mysore and BERLIN	7/14/2021 2:47:49 PM	2	en	RT @HDataSystems: Artific #hdatasystems #Artificia…
	7/14/2021 2:47:33 PM	246	en	RT @adgpi: Army Technolo
	7/14/2021 2:47:35 PM	1	en	RT @pacorjo: According to
Internet	7/14/2021 2:47:54 PM	20	en	RT @HarbRimah: Making A #MachineLearning #DataSc
	7/14/2021 2:48:35 PM	1	en	RT @weblineglobal: The ap
	7/14/2021 2:48:36 PM	1	en	RT @sokoworlddotcom: W
	7/14/2021 2:48:37 PM	1	en	RT @SuriyaSubraman: US F

Figure 3.5 – Data imported from JSON file

This shows that we already have the right columns and the right data types configured per column in this table. When importing data from a JSON file, these steps to fix the structure of the table to be able to use semi-structured data in Power BI have already been applied for you. You can see the work that Power BI has done for you when looking at **APPLIED STEPS** as shown in *Figure 3.6*:

Figure 3.6 – Query Settings pane for imported data from JSON file

As you can see, after connecting to the source, Power BI transforms the column to make sure each item or record maps to a row in the table. The table is then expanded to include the columns that map to the keys in the items. And finally, the data type for each column is changed to whatever seems most logical when looking at the data. So, AuthorLocation is identified as **Text** and CreatedAt is identified as **Date/Time** by Power BI.

If needed, you can change the columns you want to expand by clicking on the **Settings** icon next to **Expanded Column1**, as seen in *Figure 3.6*. Or you can change the data type by clicking on the icon on the left of a column header and selecting the data type you prefer to work with. And, as always, you can change the column headers to make sure the field names are intuitive for your audience.

Every second, a new tweet may appear on Twitter that includes the term *artificial intelligence*. If we want to create a near real-time report that gets new data every 5 minutes, for example, we need to work with a cloud storage solution such as Azure Cosmos DB. Many databases exist that are designed for storing semi-structured data from web applications, with Cosmos DB being one example.

Let's therefore also explore how to work with semi-structured data from Cosmos DB in Power BI. The following steps are for illustration purposes only. If you want to set this up yourself, find the full instructions on GitHub here: `https://github.com/PacktPublishing/Artificial-Intelligence-with-Power-BI/blob/main/Chapter03/set-up-cosmos-db.md`.

Now, let's see how to get data from Cosmos DB into Power BI:

1. In **Power Query Editor**, select **New Source**.
2. Select **More...**.
3. Select **Azure Cosmos DB**.
4. Provide the **URL**, **Database**, and **Collection** names. Optionally, you can provide a SQL statement here to only import certain columns or rows, for example.
5. Provide the key to authorize Power BI to import data from **Azure Cosmos DB**.
6. When previewing the data, it is very likely that a table will appear with only one column named `Document` where each row shows `Record`. In the next figure, we see a screenshot of the imported table:

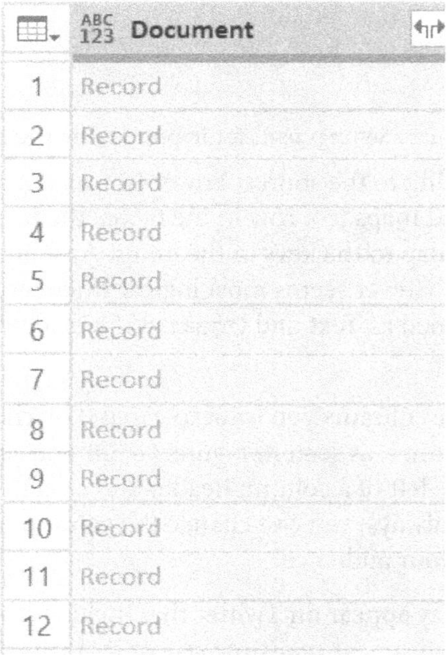

	ABC 123 Document
1	Record
2	Record
3	Record
4	Record
5	Record
6	Record
7	Record
8	Record
9	Record
10	Record
11	Record
12	Record

Figure 3.7 – Table imported from Azure Cosmos DB

This table shows data imported from Azure Cosmos DB. We only have one column, and each row represents an item or document in our database. Each of these records will have multiple key-value pairs, which is how we organize the data. To work with this data, we want the keys to become the columns and the values to fill the rows in our table. Let's try and do this.

7. To the right of the Document column header, click on the icon with the two arrows pointing left and right. This icon will allow us to expand the column. In the next screenshot, we see the pop-up menu that appears after clicking on this icon:

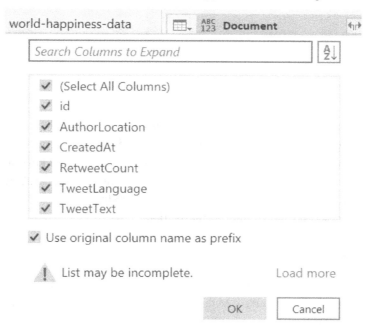

Figure 3.8 – Menu to expand columns

This screenshot shows that there are six columns that we can expand, created based on the six keys we have in the documents in our database. You can choose which columns you want to include in your table. In this case, we want all of them.

8. Uncheck the box for **Use original column name as prefix**. Select **OK** to expand all columns.

In the next figure, you can see the table that is created. The Document column has disappeared. Now the table contains the id, AuthorLocation, CreatedAt, RetweetCount, TweetLanguage, and TweetText columns.

This shows the first four columns and seven rows of the table we have created from the documents imported from Azure Cosmos DB:

	$^{ABC}_{123}$ id	$^{ABC}_{123}$ AuthorLocation	$^{ABC}_{123}$ CreatedAt	$^{ABC}_{123}$ RetweetCount
1	1415291904850153474		2021-07-14T12:47:39.000Z	0
2	1415291947560828933	Mysore and BERLIN	2021-07-14T12:47:49.000Z	2
3	1415291877897605120		2021-07-14T12:47:33.000Z	246
4	1415291886860967940		2021-07-14T12:47:35.000Z	1
5	1415291968700264450	Internet	2021-07-14T12:47:54.000Z	20
6	1415292139941109763		2021-07-14T12:48:35.000Z	1
7	1415292142256365573		2021-07-14T12:48:36.000Z	1

Figure 3.9 – Expanded table with data from Azure Cosmos DB

The column headers are based on the keys used, and the rows represent the values. As you already see in the table, not all documents have the same key-value pairs. You can see, for example, that the first row has no value for `AuthorLocation`. This is typical when working with semi-structured data because it is expected that not all documents contain exactly the same data.

To fix the structure of your data, there are the three important things we also talked about earlier when working with structured data:

- *Having the right column headers*: In this case, the column headers are based on the keys used in the data source. Make sure they are easily understandable and as a best practice, we want to avoid conjoined words. So, rename `AuthorLocation` to `Author Location` and `CreatedAt` to `Created At`.

- *Having the right columns*: You can choose the right columns when connecting to Azure Cosmos DB by using a SQL statement as described in *step 4* of the previous example. Or you can choose which columns you want to include when expanding the table as in *step 7*. Optionally, you can split or merge columns, depending on how you want to work with your data.

- *Having the right data type specified per column*: You will have to make sure you change the data type for each column. Azure Cosmos DB does not use many different data types to store data, so you will have to set it for each column when importing data into Power BI. For example, `Created At` can be set to **Date/Time**, `Retweet Count` can be set to **Whole Number,** and `Tweet Text` should be configured as **Text**. You can change the data type by clicking on the icon to the left of a column header and selecting the right data type. As a result, you can end up with a table as shown here:

ABC Author Location	Created At	123 Retweet Count	ABC Tweet Language	ABC Tweet Text
	7/14/2021 2:47:39 PM	0	en	@____bruvteresa_ Accordin
Mysore and BERLIN	7/14/2021 2:47:49 PM	2	en	RT @HDataSystems: Artificia
				#hdatasystems #Artificia...
	7/14/2021 2:47:33 PM	246	en	RT @adgpi: Army Technolog
	7/14/2021 2:47:35 PM	1	en	RT @pacorjo: According to ɛ
Internet	7/14/2021 2:47:54 PM	20	en	RT @HarbRimah: Making AI
				#MachineLearning #DataSci
	7/14/2021 2:48:35 PM	1	en	RT @weblineglobal: The apր
	7/14/2021 2:48:36 PM	1	en	RT @sokoworlddotcom: We
	7/14/2021 2:48:37 PM	1	en	RT @SuriyaSubraman: US Fɛ
In your book	7/14/2021 3:48:37 PM	2	en	RT @DD_FaFa_: Prediction I

Figure 3.10 – Table of semi-structured data with fixed column headers and data types

In this table, we have fixed the column headers to include spacing and selected the right data type for each column.

We have seen how we can work with semi-structured data in Power BI, whether that data comes from a JSON file or a cloud database such as Azure Cosmos DB. As we have seen, we need to specify in the Power Query Editor which columns we want to use and how they should be named and configure the data type for each column. Once we have checked for these three things, we have a structured table that we can work with in Power BI.

Fixing the structure when working with images

You may not think about bringing unstructured data such as plain text or images into your Power BI reports as data. Thanks to the AI features we will discuss in this book, we can use this kind of data and extract useful insights from them. In the previous section, we looked at how we can include tweet texts in our Power BI tables. In this section, let's look at how we can work with unstructured data, such as images. You may have used images in reports before, where you import the image when working in Power BI Report Builder. Here, we will look at how, in the Power Query Editor, we can import data that contains images so that we can extract insights from these images.

Before we look into how we can import images, there are a couple of considerations we need to take into account. First of all, the images you want to use should be in one of the following file formats: `.bmp`, `.jpg`, `.jpeg`, `.gif`, `.png`, or `.svg`. Secondly, you will refer to the location of your saved images and these URLs need to be anonymously accessible. See `https://docs.microsoft.com/en-us/power-bi/create-reports/power-bi-images-tables#considerations` for more information.

Globally, there are three data sources that work well when you want to import data: **Azure Storage account**, **SharePoint**, and **OneDrive**. Let's take a personal OneDrive as an example, as it is easy to set up. Make sure you have an Outlook account so you can access the OneDrive service through that account. Set up OneDrive on your system. Save the 10 example images for this exercise in a folder on your OneDrive; you can download the images here: `https://github.com/PacktPublishing/Artificial-Intelligence-with-Power-BI/tree/main/Chapter03/images`:

1.   In **Power Query Editor**, select **New Source**.

2.   Select **More…**.

3.   Select **Folder** and **Connect**.

4.   Select **Browse…**.

5.   Choose the folder on your OneDrive where you saved the 10 images and select **OK**.

6.   Select **Transform Data**.

You will get a table with the metadata of the images, where each row in the table represents an image in the folder you connected to as seen in the following figure:

	Content	Name	Extension	Date accessed	Date modified	Date created	Attributes	Folder Path
1	Binary	image01.jpg	.jpg	8/6/2021 15:44:43	8/6/2021 15:17:43	8/6/2021 15:44:43	Record	C:\Users\madie
2	Binary	image02.jpg	.jpg	8/6/2021 15:44:43	8/6/2021 15:17:52	8/6/2021 15:44:43	Record	C:\Users\madie
3	Binary	image03.jpg	.jpg	8/6/2021 15:44:43	8/6/2021 15:17:35	8/6/2021 15:44:43	Record	C:\Users\madie
4	Binary	image04.jpg	.jpg	8/6/2021 15:44:43	8/6/2021 15:17:39	8/6/2021 15:44:43	Record	C:\Users\madie
5	Binary	image05.jpg	.jpg	8/6/2021 15:44:43	8/6/2021 15:17:42	8/6/2021 15:44:43	Record	C:\Users\madie
6	Binary	image06.jpg	.jpg	8/6/2021 15:44:43	8/6/2021 15:18:09	8/6/2021 15:44:43	Record	C:\Users\madie
7	Binary	image07.jpg	.jpg	8/6/2021 15:44:43	8/6/2021 15:17:54	8/6/2021 15:44:43	Record	C:\Users\madie
8	Binary	image08.jpg	.jpg	8/6/2021 15:44:43	8/6/2021 15:17:49	8/6/2021 15:44:43	Record	C:\Users\madie
9	Binary	image09.jpg	.jpg	8/6/2021 15:44:43	8/6/2021 15:17:26	8/6/2021 15:44:43	Record	C:\Users\madie
10	Binary	image10.jpg	.jpg	8/6/2021 15:44:43	8/6/2021 15:17:47	8/6/2021 15:44:43	Record	C:\Users\madie

Figure 3.11 – Table referring to images

As you can see in this table, there are seven columns that contain information on our images that we are most likely to use (**Content, Name, Extension, Date accessed, Date modified, Date created,** and **Folder Path**) and one **Attributes** column, which you can expand for any additional metadata you want to add to your table.

The most important columns you will want to include when working with images and AI are **Content** and/or **Folder Path**. The reason is that whenever we provide an image as input for a computer vision model to predict tags or recognized objects, we need to either use the image's content or URL. For most AI services that we will cover in this book, either can be used, which is why it is good to include both in your table. The rest of the columns in your table will depend on what you want to do with this data. Do you, for example, want to plot which objects are detected in images and at which point in time? Then you would probably want to include **Date created** in your table.

Images can have an impact on the performance of your reports because they require a lot of storage space. Therefore, make sure you are careful with the number of images you refer to and the additional metadata you include in your table. It is a best practice to only include data in Power BI that you will actually use in your reports and to delete any columns or rows that you will not work with.

We have now looked at how we can import data that is structured, semi-structured, and unstructured. We have seen how we can make sure that we can work with data that is already formatted as a table in the data source, with data that is formatted as key-value pairs, and images that we want to refer to in a table containing the metadata of those images. Now that our table structures are fixed, let's look at the contents and see how we can clean our data.

Working with missing data

There are several problems you need to fix to ensure good-quality data. One of those problems is the case of missing data. This is because missing data gets in the way of us having a representative dataset and it may result in an incomplete view of reality. For many different reasons, we may have ended up with some empty rows. This may have happened when collecting, storing, or migrating the data. First and foremost, it is good to try and find out what has caused data to be missing and to see whether it can be fixed at the source. However, sometimes we just have to accept that there is data that we cannot retrieve anymore, and then questions remain regarding how we can find it and what to do with it.

Before we get into how to work with missing data, it is good to understand why we need to fix it. The problem with missing data is that it gives wrong results. For example, if we want to look at simple summary statistics, missing data can fool us. It may seem like we have only one distinct value in a column, but this could also mean that all our rows have no data and are, therefore, the only value Power BI can find. It can also negatively influence calculations done with **Data Analysis Expressions** (**DAXs**) used in Power BI. If we want to calculate the average based on the sum of all rows and divided by the number of rows we have, we will get a wrong result because an empty cell is counted to divide the sum with but it is not contributing to the sum. In other words, missing data can result in corrupted insights, which is why we need to fix it.

How do you find missing data?

In *Chapter 2*, *Exploring Data in Power BI*, we learned how to find missing data, and how we can use the **data profiling tools** in the Power Query Editor to find out whether there are values in our columns that are missing or that have resulted in an error. Remember that *errors* are often caused by setting the wrong data type on the column level or are an indication that something went wrong with loading in the data. Missing data, however, often means we did not collect that data. We, therefore, need a different solution to missing data since it does not actually exist.

Missing data can be visible as an empty cell or a cell containing the word *null* in your Power BI dataset. Sometimes it is also noted as *NaN*, which stands for *Not a Number*, and is only used when the data type of the column is, as you may have guessed, numerical.

What do you do with missing data?

In general, there are three ways to handle missing data:

- *Delete the whole row* where one or more columns have missing values.

  When deleting the whole row, you are not just deleting empty cells, you are also deleting data that you have. This is a radical move and can again result in a more incomplete dataset. Therefore, this should only be done if you have a huge dataset, and you want to remove the holes in your data to make sure that all your calculations and models will still be accurate. Although it does delete data that you could otherwise potentially use, deleting the whole row is often the easiest option. It also means that you will not have to predict or impute new values that are, at best, estimations of what the real value would have been.

- Create a model to *predict* the missing values.

  Based on trends that a model may recognize and learn in your data, you can get an estimation of what the missing value should have been. This is mostly beneficial if you have only a few columns in which you have missing values and you can use your other columns to train a model. We can train a model on our data in Azure Machine Learning, and integrate that model with Power BI to generate the predictions in Power BI itself. The technical details of this will be covered later in *Chapter 12, Training a Model with Azure Machine Learning*.

- *Impute* the missing values with some simple statistical metrics, such as the **mean** or **median**.

  This is a method very often used when deleting a whole row and when training a model is not an option. Just like when using a model, you are replacing the missing values with estimations; only this time, the estimations are not based on some pattern a model has learned to recognize. Instead, we are using simple averages, or middle values, to fill the holes caused by missing data.

  This method is called **imputing missing values** and is most often done with means or medians. However, you can choose another value altogether if that makes sense for your data. The reason why means and medians are often used is that they will affect your averages and calculations the least. So, it is great if you are mostly concerned with global insights and less interested in individual values. This is why in data science, we often use this method to prepare the data for when we want to use it as input to train a model.

We now know how we can recognize, and work with, missing data. We can either delete rows in our table that have missing data, train models to complete our dataset, or use statistical metrics such as means and medians to impute missing values. Next, we will look at what we can do when we have complete data that is imbalanced.

Mitigating bias

It is important to know whether you have a biased dataset, as it may mean you do not have a representative dataset. It may also mean that you will produce an AI model that treats different groups unfairly. For example, your AI model accurately forecasts supermarket sales in cities but underperforms in towns. If you use demand forecasting to plan the supplies to send to each supermarket, this can result in constant supply shortages in your supermarkets that are in towns. (We'll talk about forecasting in *Chapter 4, Forecasting Time-Series Data*.)

When we talk of **bias**, we often refer to imbalanced data. To know whether a dataset is imbalanced, we mostly look at histograms, box plots, and the distribution of values. As soon as we see that there is an inequality in the number of values we can find in our dataset, this can mean that there is bias in our data.

Bias is a complex problem and does not have one golden solution. When your data is biased, understanding the cause of the bias will help to decide how to fix it. We will first discuss how to find bias, and then we will discuss what to do with it. We'll go into more detail on the importance of mitigating bias in *Chapter 13, Responsible AI*.

How to find bias

There are different types of bias. The most straightforward feature of bias is that you have an *imbalanced* dataset. This imbalance can mean, for example, that of 100 participants, 10 were women and 90 were men. As soon as one group is less than 20%, we often talk about a *minority*, and we need to be careful with these classes as they can be neglected when we train an AI model. Remember the *profiling* tools covered in *Chapter 2, Data Exploration*, which will help you find these kinds of imbalances.

To correctly identify bias, you must be aware of the different causes of it. First of all, bias can be caused by how data is collected, which is why you always have to understand how the data has been created and what it represents. If, for example, you want to get insights into how fast someone can run 5,000 meters and you only measure the running times of top athletes, you may have a biased view. In this case, it is good to understand your goal. Is it to know the average time of any person or that of a top athlete? If needed, collect more data; also measure running times of people who do not train to run.

If you believe you collected the right data and that you do have a representative dataset, you can still simply have a skewed distribution in your data. Look at the histogram of *Healthy life expectancy at birth* for the *World Happiness Report* dataset in the following figure:

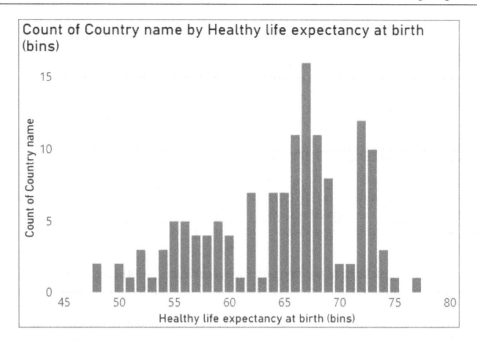

Figure 3.12 – Left-skewed distribution for Healthy life expectancy at birth

(See *Chapter 2, Exploring Data in Power BI,* for more information on how to produce a histogram.)

As we can see in the histogram in *Figure 3.12*, there is an imbalance. There are more countries where the healthy life expectancy at birth is higher than 65, resulting in a left-skewed distribution. This could very well be reflecting reality. In that case, it is good to be aware of this imbalance, but there may not be a need to fix it.

Another interesting case is when you know you have a representative dataset, and there is a bias, but you do not want that bias to be there. For example, according to Statistics Netherlands (https://www.cbs.nl/nl-nl/visualisaties/ dashboard-arbeidsmarkt/banen-werkgelegenheid/toelichtingen/ werkgelegenheidsstructuur), of all the people working in healthcare in 2020, 80.9% were female and 19.1% were male. This again could be true and could be the result of cultural bias. In this case, we know that there is an imbalance because we have two categories, or classes, with an unequal distribution: female and male, of which one is less than 20% and, therefore, a minority.

Whether this is a problem depends on what you do with the data. If you are the manager at a hospital and simply want to visualize these insights to understand the demographics of your employees, it does not necessarily have to be a problem. If you are a hiring manager and want to train a model to predict who to hire based on personal information including sex, then there is a large chance the model will reinforce the bias and think that only women can work in healthcare. This is when you want to fix the bias in your data.

How to mitigate bias in your dataset

If you have a biased dataset and you want to use it for training an AI model, there are a couple of things that you can do. The first option focuses on the data: *resampling* your dataset. If you consider the rows in your table to represent the samples in your dataset, the purpose here is to make sure you have a more balanced and equal amount of samples per class.

For example, if we think back to women and men working in healthcare, we want an equal amount of women and men in our dataset when we train a model, even if that is not the distribution we see in reality. If you have a large dataset, you can simply *under-sample* and remove samples or rows containing women, until you have a balanced dataset. This means you are deleting data, so should only be done if you think you will be left with enough data to train a model.

Alternatively, you can *over-sample* and add new samples of men to make sure you end up with an equal amount of women and men in your dataset. The question then is, how do you add data? An important side note here is that sometimes, this can simply be fixed by collecting more data but then focusing on this one underrepresented class in your dataset. But, assuming that is not possible, we would have to *create* new samples. A straightforward approach to this would be to add samples by duplicating data.

This is not always considered the prettiest approach, which is why there are some other techniques available in data science to help create this new data. For example, the **Synthetic Minority Over-Sampling Technique** (**SMOTE**) can be used where an algorithm will add new samples for you by looking at the existing data. You will mostly use this when training a model yourself, which is why we do not see an option for this in Power BI. However, we will see how we can use this technique in *Chapter 12, Training a Model with Azure Machine Learning*.

Lastly, bias is not by definition a problem in your dataset. Sometimes the fix to this problem is simply selecting a different algorithm to train your model with. Again, this depends on what your goal is. If you know that you have a minority class in your dataset and that is exactly what you want to train your model on, you may just have to choose the right type of machine learning algorithm, such as **anomaly detection**. Imagine you work for a bank, and you want to identify when a transaction is fraudulent. Hopefully, this only occurs for less than 20% of all transactions, making it a minority, and you want to train a model to recognize exactly these transactions. You can then leave the expected bias in your data and train an anomaly detection model to make the model aware that these samples are rare and should be compared to the normal in order to be detected. We will talk more about this in *Chapter 5, Detecting Anomalies in Your Data Using Power BI*.

Bias can exist in your dataset because of how you collected your data, because of expected imbalance, or because of cultural bias. First and foremost, we need to recognize it and understand its cause. Then, we need to have a clear goal of what we want to do with our data to know what we can do to best mitigate bias. This can mean we have to resample our dataset, or it can mean we have to train a different type of model. Whatever you choose to do, the main purpose should be to create fair and accurate models.

Handling outliers

When we talk about outliers, we are referring to those observations that are very different from the rest of our data. Sometimes, outliers are exactly what we are looking for, such as when we want to detect anomalies in a running engine, or when we want to detect fraudulent transactions. Other times, outliers are mistakes in data collection and can result in a less accurate model. It is important to know whether you have outliers in your dataset, know what they represent, and remove them if necessary.

The common approach to finding outliers is by using a box plot. In *Chapter 2, Exploring Data in Power BI*, we created one for the **Life Ladder** score in 2019 of the *World Happiness Report* dataset as seen in the following figure:

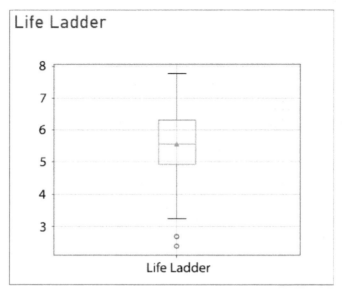

Figure 3.13 – Box plot of Life Ladder including outliers

In this figure, the box plot shows the distribution of the **Life Ladder** scores for all countries. At first glance, it seems to be normally distributed, so nothing is wrong. If you choose to show outliers, then you will see two dots at the bottom of the box plot. These two dots represent two extremely low values for **Life Ladder**. Upon further inspection, we can see that these two dots represent Afghanistan with a score of 2.38 and Zimbabwe with a score of 2.69. In this case, these are low but realistic values, so it is good to check for them but there is no need to *fix* them.

When does an outlier become a reason for concern? Let's look at another example from the *World Happiness Report* dataset. This time, we have a box plot of *Healthy life expectancy at birth* in 2019 in the following figure:

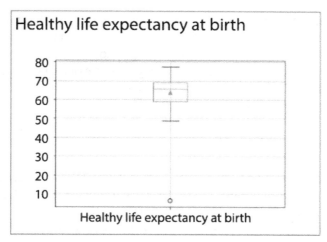

Figure 3.14 – Box plot of healthy life expectancy at birth

In the box plot here, we see that most values are roughly between 50 and 80. There is one outlier that is smaller than 10. This is an extremely low value and seems unrealistic. The most likely explanation for such an outlier is that something went wrong when collecting data. Instead of 67, someone may have accidentally put in 6.7. If we can find out what happened, and verify what the true value should be, we can change this value to 67. However, if you cannot check what the cause of this value is and it seems unrealistic or you expect it to confuse your model, it is probably best to simply delete this observation. This means you will probably have to delete the entire row (although you could treat it as missing data and use one of the techniques for handling that).

As a final note, as mentioned earlier, sometimes outliers are exactly what we are looking for. It is always a best practice to check for them and understand them. It could be that it is the outliers that you want to find because they represent the rare events that you want to be able to predict. For example, if you create a Power BI dashboard because you want to monitor the temperature of a car engine so you know when it needs maintenance, you expect the temperature to fluctuate within a certain range. As soon as you measure extremely high temperatures in your car engine, these can be considered outliers. Nevertheless, those temperature measurements are crucial as they can indicate that your engine is in a dangerous state and should be checked immediately. In this case, outliers should not be deleted and instead should be included when training an anomaly detection model to be able to predict these events in the future.

Besides looking at the distribution of your data by plotting a box plot, it is always good to check for outliers with this visualization. Knowing whether you have outliers and finding out what causes them ensures that you can react adeptly to them. If you think outliers will confuse your model, you can delete them. If you expect these outliers, you may also train a model to recognize them. Either way, it is another characteristic of your dataset that helps you get to know it from top to bottom.

Summary

In this chapter, we have worked on the exploration of structured, semi-structured, and unstructured data in the Power Query Editor. We have imported text and images and have made sure that Power BI knows how to interpret this data so that we can later use AI features to extract insights from this rich data. We have looked at the three common problems with data when it is used for AI: missing data, bias, and outliers. We have discussed the different considerations for each of these problems where it is important to understand what causes them and whether they will cause problems for what you want to do with your data. We also covered how we can fix missing data, bias, and outliers, if needed, to make sure we have a representative dataset that will result in building accurate models.

In the next chapter, we will discover the first type of model we can work with in Power BI: forecasting.

Part 2:
Out-of-the-Box
AI Features

Without needing any data science knowledge, in this part, you'll learn how to use out-of-the-box AI features already available in Power BI.

This section includes the following chapters:

4

Forecasting Time-Series Data

How can we make predictions about the future? We want to make data-driven decisions on how to move forward. To know what to do next, we will need some idea of what will happen in the future, based on what has happened in the past.

Forecasting is a form of machine learning that uses historical data to predict what will happen in the future. A quick and easy option to generate forecasts about your data is by using the out-of-the-box option in Power BI. We can generate forecasts on any time-series data. For example, we can visualize the expected product sales for the coming months, based on historical data.

In this chapter, we'll learn how we can make use of one of the easiest and most fruitful machine learning applications in Power BI: forecasting time-series data. As forecasting is an out-of-the-box option, it is very quick to add to your reports. To make sure the results make sense, however, we do need to go over the data requirements and understand the logic behind the model before we can actually use the feature.

We're going to cover the following topics:

- Getting the data right for forecasting

- Understanding how forecasting works

- Using forecasting in Power BI

Let's begin by looking at the prerequisites of this chapter.

Technical requirements

There are two things you need to walk through the examples provided in this chapter.

- **Power BI Desktop**: As this book revolves around Power BI, we expect you to have Power BI Desktop installed on your system. You can install Power BI from the Microsoft Store or find more advanced downloading options here: `https://www.microsoft.com/en-us/download/details.aspx?id=58494`.

- **A sample dataset**: A sample dataset of the amount of tourists per month in the Netherlands is used in the examples provided in this chapter. The data is collected by *Statistics Netherlands* and can be found in their online open database here: `https://opendata.cbs.nl/#/CBS/nl/dataset/82058NED/table?searchKeywords=logiesaccommodaties`. The cleaned-up data that is used in the examples throughout this chapter can be found on GitHub: `https://raw.githubusercontent.com/PacktPublishing/Artificial-Intelligence-with-Power-BI/main/Chapter04/tourism-data.csv`.

Data requirements for forecasting

To integrate forecasting in your visuals in Power BI, there is one thing you need: data. The option to add a forecast will not even appear in Power BI if your data does not meet the necessary requirements. As well as that, the volume and quality of data can influence the accuracy of the forecast that Power BI will create for you.

First, let's learn a little bit more about forecasting. Then, we can explore the data requirements for forecasting in Power BI and illustrate it with an example.

Why use forecasting?

In general, the purpose of forecasting is *to make predictions on what will happen in the future*. There can be a myriad of reasons why anyone would want to know what will happen in the future. If we keep it within the scope of business intelligence, it is highly likely that forecasting is used to form actionable insights. For example, if we keep on doing business as usual, how many products should we expect to sell next year? Or, based on trends that we observed over the last few months, what will happen to the housing prices in the next month?

Having expectations of what will happen in the future can help us to make business decisions. If we expect to sell a certain amount of products in a year but we see that the current sales are lower than what we expected, we may have to invest more in marketing to make sure we keep on growing as a business. And if we expect housing prices to go up in the next month, a bank may want to rethink the interest rate for mortgages.

The approaches to forecasting differ in their complexities and will depend on your use case. A *simple* approach is to run statistical analysis of the target variable over the historical data to infer what future data will look like. For example, your product sales have been consistently going up in the past, and you expect them to grow in the same way in the future. This approach does not take into account any external factors that may influence the target variable. For example, you are not looking at the effect of competitor activity on your product sales.

A more *complex* approach is to see how other features have influenced the target variable in the past and use this information to potentially make more accurate predictions of the future. In this case, you would be taking into account competitor activity and its effect on your product sales. These external factors can, however, be hard to define and require a lot of proper data analysis to make sure that they are included correctly in your forecasting models.

In this chapter, we are focusing on the low-hanging fruit; we want to explore the easy approach to forecasting that can be attained by using the built-in feature in Power BI. This feature uses the simpler approach of only looking at the target variable's historical data to generate predictions. We are not looking at the influence of other factors when using this feature yet. More complex forecasting models can be created with Azure Machine Learning and automated machine learning, which will be discussed in *Chapter 11, Using Automated Machine Learning with Azure and Power BI*.

Since Power BI takes the simple approach to forecasting and will only look at the target variable itself, there is not much data we need besides that. The only other variable that we need to include in the data is time, which we will discuss next.

Time-series data

When we create a forecasting model, the model learns from historical data to create predictions about the future. The most common type of data used for forecasting models is commonly referred to as **time-series data**, as we have data *sequentially* collected over a *series* of *time* data points.

Time-series data needs to meet the following requirements:

- A **target variable**: Include a target variable that is *numerical* – what it is that needs to be forecasted based on historical data? This can be something such as product sales, house pricing, energy consumption, or the amount of tweets mentioning your brand. It can be any numerical variable that you want.

- A **time variable**: Include a time variable that is a *date*, a *date/time*, or a uniformly increasing *whole number* – the target variable should be measured over a regular interval of time. For example, the amount of tweets mentioning your brand per day, the energy consumption of your office building per week, or the product sales per month.

> **Minimal Missing Data**
>
> Models designed to forecast data do not like gaps in historical data and will impute missing values. Imputed values are not observed data points but are computer-generated values, similar to observed values, for filling the gaps. To avoid these imputed values corrupting your model, it is best to identify whether you have missing data and fix this yourself before forecasting. Fixing missing data is covered in *Chapter 3, Data Preparation*. The fewer missing values, the more accurate your model is likely to be.

To illustrate the two variables needed for forecasting, we'll look at a simple example dataset that contains these two fields.

Using an example – tourism data

The Netherlands welcomes many tourists each year. The number of tourists visiting the country is measured by Statistics Netherlands every month. In this case, the target variable is the amount of tourists, and the time variable will include an entry for each month from 2012 to 2020. In *Figure 4.1*, we can see the first 10 rows of this data, which have been imported into Power BI:

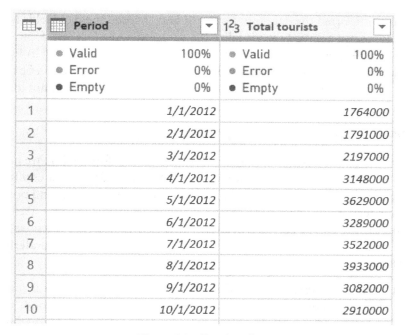

	Period	1²₃ Total tourists
	● Valid 100%	● Valid 100%
	● Error 0%	● Error 0%
	● Empty 0%	● Empty 0%
1	1/1/2012	1764000
2	2/1/2012	1791000
3	3/1/2012	2197000
4	4/1/2012	3148000
5	5/1/2012	3629000
6	6/1/2012	3289000
7	7/1/2012	3522000
8	8/1/2012	3933000
9	9/1/2012	3082000
10	10/1/2012	2910000

Figure 4.1 – Tourism data

As you can see in the figure, **Period** is a date field, and **Total tourists** is a whole number field.

If you want to follow this example, follow these instructions to import the data into Power BI:

1. Import the data by downloading the dataset from here: `https://raw.` `githubusercontent.com/PacktPublishing/Artificial-` `Intelligence-with-Power-BI/main/Chapter04/tourism-data.csv`.

2. Save the CSV file somewhere on your local system.

3. We want to work with the date field as a hierarchy, so enable **Auto Date/Time** for **Report Data Load**.

4. Import the data into Power BI. If you need guidance on how to import the data, see *Chapter 2, Exploring Data in Power BI*.

5. Open the Power Query Editor.

6. It is highly likely that Power BI will autodetect the correct data types. To change the data type of a column or field, you can select the icon on the left of the column header. This way, you can also change the locale to make sure that the formatting of the **Date** column corresponds with the language you use.

Now that we have imported the data into Power BI and checked the data fields for the necessary data types, we are ready to visualize the data and add a forecast. However, it is good to dive a little deeper into what is happening when we use forecasting in Power BI before we actually use it. By understanding the algorithms behind it, we will be more able to judge when to use the forecasting feature and, more importantly, when not to.

Algorithms used for forecasting

To properly make use of a machine learning technique such as forecasting, we need to understand how it works. Just making sure we have time-series data is not enough to get accurate predictions in Power BI. We need to understand the limitations and possibilities of the forecasting feature in Power BI to ensure that our results make sense. We'll first discuss how forecasting can be calculated, after which we will describe when we should use this feature in Power BI, based on the limitations and possibilities the underlying technique gives us.

The reason why some are skeptical of using an out-of-the-box feature such as forecasting in Power BI is because it is unclear what is actually happening under the hood. This kind of feature is often described as a black-box model – we can't see how the model is trained and how it creates predictions. Unfortunately, this is the intellectual property of Power BI and Microsoft and is, therefore, not made public. However, we can still guess what is happening behind the scenes and make sure that we are aware of the limitations and possibilities when using this forecasting feature in Power BI.

The benefit of using an out-of-the-box feature

Before we can explore how forecasting in Power BI is calculated, we need to be aware of what it means for this feature to be called *out of the box*. We can also refer to such a feature as a *black box*, as there is no transparency offered to how exactly the predictions are calculated.

This may sound scary and will make it less interesting for many users to actually add this feature to their reports. There is, however, an upside to such a feature: you save a lot of time, effort, and expertise that you need to have in-house. *Figure 4.2* shows a trade-off between the downside and upside of using an out-of-the-box feature:

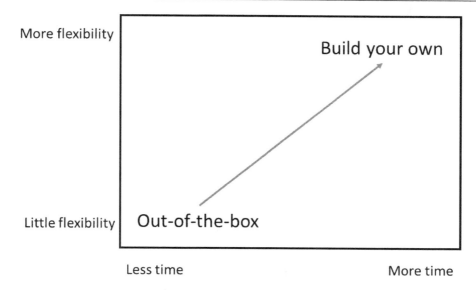

Figure 4.2 – The trade-off between an out-of-the-box feature and building your own

As you can see in the diagram, the trade-off is fairly simple. You can invest very little time to use an out-of-the-box model, which means you have little flexibility in tweaking the model exactly to your needs. If you invest more time to actually build your own model, you can have more control over the model and fine-tune it to your data.

In an ideal world, you will probably want to create your own models. However, we don't always have the resources, nor does it always yield much better results. More complex and customized models don't necessarily have to be more accurate. However, when using an out-of-the-box model, we do have to venture a guess at how it actually works so that we know how to use it best. So, let's try and do that.

Understanding how forecasting is calculated in Power BI

The future is a mystery. When forecasting, we try to make predictions about the future that take into account patterns found in historical data. But what do we mean by that? To understand what a simple forecasting model can do, let's explore how we can calculate predictions about the future and how Power BI can do this intelligently.

First of all, think about why you want to use a forecasting model. One of the main purposes of a simple forecasting model is to detect two components in your data:

- **Trend**: An increase or decrease of the value of your target variable, usually over a long period of time
- **Seasonality**: A pattern that can be perceived over a fixed period of time that repeats itself

As stated earlier, Power BI's forecasting model is a simple approach to forecasting. With simple approaches, we only look at the target variable and how it varies over time. We do not take any external factors into consideration when we run the statistical analysis. What this means is that we look at whether we can detect trends or seasonality for the target variable. We then use that information to predict what may happen in the future.

For example, let's look at the tourism data for the Netherlands from 2012 until 2020, as introduced earlier in this chapter. We can create a **line chart** that plots the amount of **total tourists** as **values** and the **period** as the **axis**. This will result in a chart such as the one shown in *Figure 4.3*:

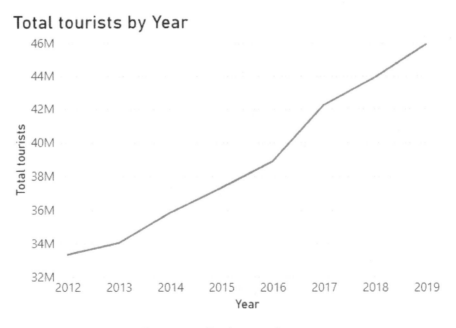

Figure 4.3 – Total tourists by year

As you can deduct from this figure, the amount of tourists that have been visiting the Netherlands each year has been steadily going up. This increase shows that there is an upward trend. A trend doesn't always need to only be upward or downward; if we add the years 2020 and 2021 to this dataset, we will actually very likely see a downward trend for these years because of the impact that COVID-19 has had on tourism. Nevertheless, this figure illustrates what we mean by a trend.

Luckily, this dataset doesn't only have the amount of tourists measured per year. We also have the amount of tourists that have visited the Netherlands each month. If we expand the data down two levels in the period hierarchy, we get a visual representation of the data per month, for each year, as seen in *Figure 4.4*:

Total tourists by Month

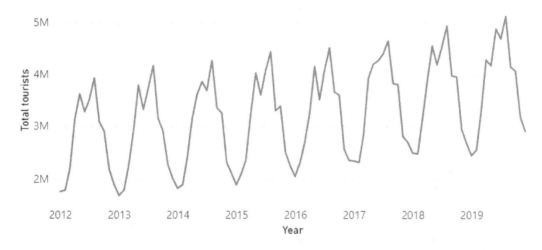

Figure 4.4 – Total tourists by month

Again, this line chart shows that there is an upward trend; the total tourists each year have increased. But now, we can also see that there is seasonality. Within each year, during a fixed and known time period, the amount of tourists starts low, goes up and down a couple of times, and ends low.

Now that we know the time period in which we have seasonality, namely a year, we can take a different perspective and plot the average seasonality over a year to understand it a bit better. For this, we drill up again to show the total tourists by year, after which we go to the month level in the hierarchy. This results in the plot shown in *Figure 4.5*:

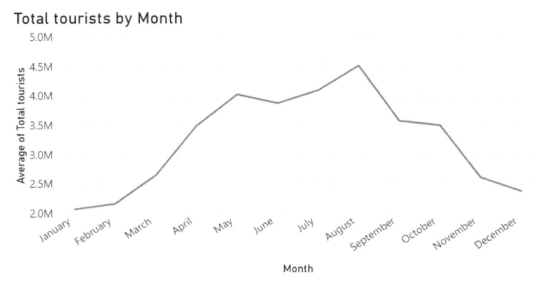

Figure 4.5 – The average of the total tourists by month

In this figure, showing the average of **Total tourists by Month**, we can explore the seasonality a bit more. Although we start with little tourism each year in January, there is a huge increase in tourism up to May, after which there is a slight dip in June, followed by the largest peak on average in August. And with the end of the summer comes the end of tourism, which drops to its lowest point again in December. For a country with summer in July and August, this seems to make sense and is in line with what you may expect.

So, for tourism in the Netherlands, we now know the following:

- **Trend**: There is an overall increase in tourism from 2012 until 2019.
- **Seasonality**: One season can be defined as one year. In one year, we can see a pattern that repeats itself every season.

One straightforward forecasting technique that doesn't take external factors into consideration and focuses on trend and seasonality is the collection of **Error, Trend, and Seasonality (ETS)** models. There is a lot of math involved to calculate the forecast using any type of ETS model, and the purpose of this chapter is not to recreate it but to understand its limits. For that reason, we will focus on the most essential information.

These models use **exponential smoothing methods**, which means that not all data points in your historical dataset are weighted the same. Instead, the more recent observations are deemed the most important, and the older the observations, the less they are used to calculate future predictions.

What the ETS models then do is detect three parts of your data:

- The average value of your target variable over the period of data you have
- The trend
- The seasonality

By detecting these three components, a model can then understand the patterns in your data and use that knowledge to estimate what will happen in the future.

As mentioned before, we don't know exactly what Power BI uses. So far, we know the following:

- The configuration settings we see when using the forecasting option in Power BI (we'll explore those in the next section), which are similar to the options we have when training an ETS model.
- That ETS models are also used in Excel (see https://support.microsoft.com/en-us/office/forecast-ets-function-15389b8b-677e-4fbd-bd95-21d464333f41).
- That using the exponential smoothing method is a very popular approach to forecasting, as it has proven to be highly successful when working only with the target variable. In Power BI, we work with the target variable only.

Based on these three facts, it seems safe to assume that Power BI might be using ETS models.

Now that we have learned about trend, seasonality, and the algorithms behind the forecasting model in Power BI, let's talk about the most important thing: how to use forecasting in Power BI properly to optimize its accuracy.

Optimizing forecasting accuracy in Power BI

Now that we understand what is important for the forecasting models used by Power BI, how can we use them to make accurate predictions for our own historical data?

Accuracy in this case refers to the predictions being very close to the real data we will observe in the future. You expect a little error, or difference, between them, but you want to minimize this. To achieve this, we need to make sure we use the forecasting feature wisely.

As with any venture into AI, before you add any intelligence to your data, you should explore your data and really know it inside and out. We discussed the *data requirements*: we need to have time-series data that has a target variable, a time variable, and (almost) no missing data.

Besides the data requirements, forecasting works best in Power BI if you see a *clear linear trend* in your historical data. You can check for a trend by adding a trend line to your line chart with the following steps:

1. Create a line chart with the **Total tourists** target variable as values and the **Year** time variable on the axis, as shown in *Figure 4.3*.

2. Select the **Analytics** tab in the **Visualizations** pane.

3. Expand the field for **Trend line**.

4. Select **+Add**.

A trend line is automatically added to your line chart. The result is shown in *Figure 4.6*:

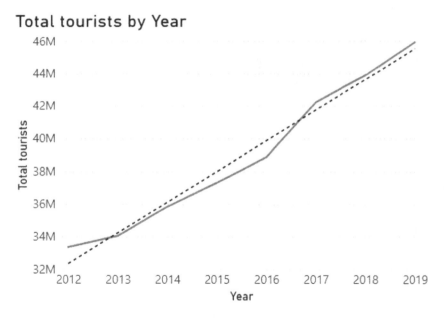

Figure 4.6 – A trend line added to the line chart

In this figure, we can see a linear upward trend that follows the data very closely. When these two lines are close to each other, it means it will be easier for Power BI to calculate the forecasts.

Let's compare that with another view where the trend is not so linear. If we include the tourism data of 2020 and the first half of 2021 in our tourism dataset, it will be a different story. If we plot the tourism data of 2012 until June 2021 in a line chart and add a trend line, we get the following figure:

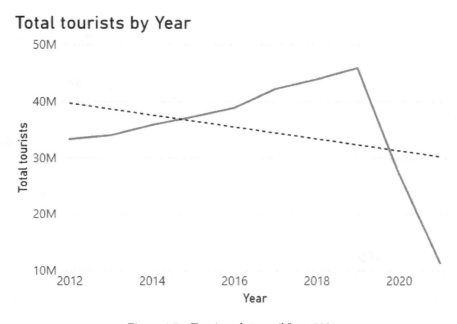

Figure 4.7 – Tourism data until June 2021

In *Figure 4.7*, the trend line does not closely follow the actual data. This is because there was a steady upward trend until the beginning of 2020, and then we had a very unexpected big drop in tourism in the Netherlands. Understanding the context, we can draw conclusions as to why this has happened. This kind of event could even be caused by *outliers*, as they differ greatly from the typical data points we collected. (Learn more about outliers in *Chapter 3, Data Preparation*.)

In this case, we can't exclude these outliers, as they are not rare and still represent a different reality from the data before 2020. The conclusion based on the non-linear trend we can see in *Figure 4.7* is that Power BI will have a hard time making an accurate forecast for this dataset. You can always try! But *proceed with caution* whenever you have data that doesn't show a linear trend when adding a forecast in Power BI.

The previous example also highlights another important consideration when forecasting with Power BI. Whenever we calculate the forecasts of a target variable without taking into consideration any external factors (and even if we do), the expectation should be that the *future is similar to the past*. If we think the future is very different from the past, then Power BI's forecasting feature may not be the right choice. Forecasting models, in general, may not even be the right choice, as they may require more strategic thinking and contextual information to make more accurate forecasts, which is something humans can be better at.

Not having a linear trend, or not expecting a similarity between the past and future, are *not* guarantees that your forecast will *not* be accurate. Instead, it means you should be careful about trusting and using the predictions. They may still reveal useful insights!

To optimize the accuracy of the forecasting models used by Power BI, you need to make sure that your data meets the requirements and contains a target variable, a time-series variable, and (almost) no missing data. The most appropriate historical data shows a linear trend that we expect to continue in the future. With this in mind, you are more likely to get accurate results.

Using forecasting in Power BI

We collected historical data and imported it into Power BI, and we considered the limitations of forecasting in Power BI. Now, let's go through the steps of adding a forecast to a line chart in Power BI and how to configure it.

Before you walk through the following steps, make sure you have imported the tourism dataset into Power BI, as described in the *Using an example – tourism data* section.

Once you have imported the dataset into Power BI, you should have a table or query named `tourism-data` that consists of two fields: `Period` (categorized as **date**) and `Total tourists` (categorized as **whole number**):

1. Create a line chart with **Year** on the axis and **Total tourists** as the values.

2. Select **Expand all down** to go down two levels in the hierarchy. You should have a line chart, as shown in *Figure 4.8*:

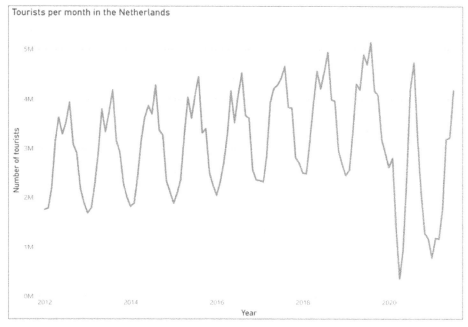

Figure 4.8 – A line chart of Total tourists by Year

This means that we visualized the amount of tourists per month from 2012 until 2019. Because we expanded the data down two levels (from year to month), we can see the seasonality in our data. The data is measured per month, and a season is measured as one year.

3. Select the **Analytics** tab in the **Visualizations** pane.

4. Expand the field for **Forecasting**.

5. Select **+Add**.

   A forecast will be added to your line chart; it may not make sense yet. So, let's explore the configuration options first.

There are four things you can configure when adding a forecast, as shown in *Figure 4.9*:

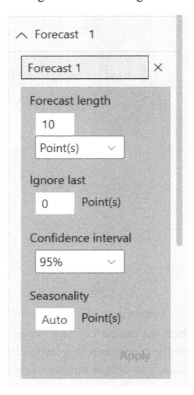

Figure 4.9 – Forecasting options

In this overview, you can see what each option means:

- **Forecast length**: How far do you want to predict in the future? You can choose how many points you want to predict in the future. One data measurement is considered one *point*. So, if you collected the amount of tourists per month, one month is seen as one point. You can also change from using points to actual time measurements, as shown in *Figure 4.10*:

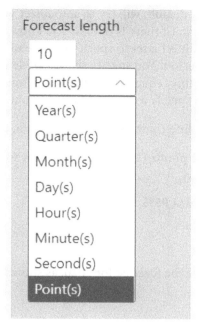

Figure 4.10 – Forecast length options

Using time measurements such as **Month(s)** instead of **Point(s)** may make more sense from a usability perspective. Use whatever you prefer.

- **Ignore last**: To verify the model's accuracy, you can specify that you want to ignore the last so-many points. Power BI will then *predict* from the point that is included last after the amount of indicated points is ignored. Then, it will predict the amount of points or time period you have specified in the forecast length. By doing this, you can see how well the forecasts compare to the actual data, which helps in deciding whether the model is accurate enough for you to use.

- **Confidence interval**: Whenever we make predictions, a model can give a specific value as a prediction, as well as a range of values within which the actual data may fall. This range of values is known as the **confidence interval** and shows in your line chart as the gray area around the forecast. Adding the confidence interval to your visual allows you to see how accurate the model is. If the gray area is narrow, the model is likely to be accurate; if the gray area is wide, the model is very unsure of its predictions, which can be a reason for concern. Increasing the confidence interval increases the range of values that you see in your visual. For stricter predictions, you can opt for a lower number.

- **Seasonality**: Power BI, by default, will try to autodetect what the fixed period is in which it can detect a season (if there is one). If you know that there is seasonality in your historical dataset, it is better to specify the size of one season to ensure that Power BI uses the right period. You specify the season length in *points*. For example, if you collected data monthly (equal to one point) and one season is 12 months or one year, then you would specify a seasonality of 12 points.

After adding the forecast to the line chart for the tourism data, configure the following:

1. Set **Forecast length** to 12 points (12 months or 1 year would give the same result).
2. Keep **Ignore last** at 0 points.
3. Keep **Confidence interval** at **95%**.
4. Set **Seasonality** at 12 points.
5. Select **Apply**.

   As a result, you should see the forecast for 2020, as shown in *Figure 4.11*:

Total tourists by Year

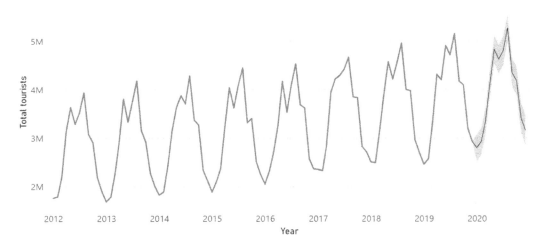

Figure 4.11 – A forecast for the total tourists

In *Figure 4.11*, we can see that Power BI appears to be very capable of forecasting the tourism data for 2020. We see a similar seasonality in 2020 but also a continuation of the upward trend that we saw previously in our data. The gray area, or confidence interval, is very narrow around the forecast, which we can interpret as the model being fairly accurate. Unfortunately, as we now know, the model was not right, as it could not predict the unforeseen events that we experienced from that year onward. Just to see a not-so-perfect forecast, let's have a look at what happens if we take that data into account.

In *Figure 4.12*, we can see a line chart visualizing historical data on tourism from January 2012 until June 2021. At the end, a forecast is added for 18 points in the future: from July 2021 until December 2022, with a seasonality (still) of 12 points:

Total tourists until June 2021

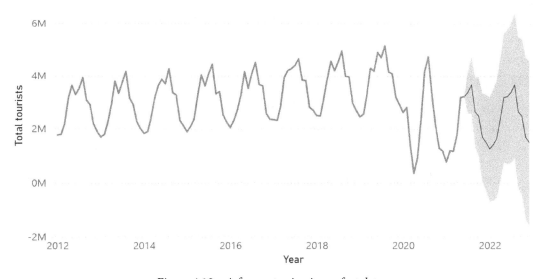

Figure 4.12 – A forecast using imperfect data

The first thing that may get your attention is the shaded area around the forecast in *Figure 4.12*, which is much wider than in *Figure 4.11*. Since the beautiful pattern that occurred in the historical data until 2019 was disrupted by 2020 and the first half of 2021, the model is much less confident about the predictions. As explained earlier, this shows that the model isn't as accurate, as the actual values may vary a lot from what is predicted, shown by the black line representing the forecast. Also, remember that this forecast does not take any external factors into consideration, so it is really up to the user to decide whether it is a useful forecast or not.

Now you know how you can add a forecast in Power BI to a line chart and, more importantly, how you can configure it to make sure the forecast represents your business needs.

Summary

In this chapter, we talked about three major topics to ensure that you are using the forecasting feature in Power BI with sense. We discussed the data that should be used, which should include target and time variables and have little to no missing data. We explored the potential algorithm behind the forecasting feature in Power BI at a high level to understand its limitations and how to use it properly. And finally, we actually added a forecast to a line chart in Power BI to see how it works in practice and what we can and should configure to create accurate forecasts. In the next chapter, we will continue the conversation on time-series data. Instead of forecasting, we will explore how we can identify anomalies that represent an unexpected change in the data.

Further reading

- *Forecasting: Principles and Practice, Rob J Hyndman and George Athanasopoulos*: `https://otexts.com/fpp3/what-can-be-forecast.html`
- *Time series Forecasting in Power BI, Sandeep Pawar*: `https://pawarbi.github.io/blog/forecasting/python/powerbi/forecasting_in_powerbi/2020/04/24/timeseries-powerbi.html#ETS(AAA)`

5
Detecting Anomalies in Your Data Using Power BI

Anomaly detection is used when unexpected and rare events need to be identified. Some anomalies are clear **outliers** and show up as easily recognizable spikes in data. However, some anomalies are more subtle than that and require **machine learning (ML)** to be detected.

What an anomaly represents depends on the situation. Most commonly, anomaly detection is used for **predictive maintenance**, when monitoring—for example—an engine. The temperature of an engine can be measured and visualized and is expected to fluctuate. However, whenever the temperature increases or decreases significantly and unexpectedly, it is a cause for concern and a reason to further investigate.

Power BI has an out-of-the-box **anomaly detection** feature that can be used to detect unexpected events in time-series data. Having this feature makes it very easy for users to find out which data points don't fit in the normal trend of the data, and it can even explain why an unexpected event took place.

As with any of the out-of-the-box **artificial intelligence** (**AI**) features in Power BI, it helps to understand how to use it and—especially—when not to use it. To make sure we use the anomaly detection feature correctly in Power BI, we will cover the following topics:

- Which data is suitable for anomaly detection?
- Understanding the logic behind anomaly detection
- Using anomaly detection in Power BI

By the end of this chapter, you will know what the requirements are of data you want to use with anomaly detection and how you can enable the feature in Power BI.

Technical requirements

There are two things you need to walk through the examples provided in this chapter, as outlined here:

- **Power BI Desktop**

    As this book revolves around Power BI, we expect you to have Power BI Desktop installed on your system. You can install Power BI from the Microsoft Store or find more advanced downloading options here: `https://www.microsoft.com/en-us/download/details.aspx?id=58494`.

- **Sample dataset**

    A sample dataset is used in the examples provided in this chapter. The data is collected by *Statistics Netherlands* and can be found in their online open database here: `https://opendata.cbs.nl/#/CBS/nl/dataset/82058NED/table?searchKeywords=logiesaccommodaties`. The cleaned-up data that is used in the examples throughout this chapter can be found on GitHub at `https://raw.githubusercontent.com/PacktPublishing/Artificial-Intelligence-with-Power-BI/main/Chapter05/tourism-details.csv`.

Which data is suitable for anomaly detection?

Anomalies are unexpected data points, so it is likely they are not even visible at first glance. There are, however, some other requirements your data needs to meet before you can make proper use of the anomaly detection feature in Power BI. Before we talk about the data requirements, let's first talk a little bit more about why we would use anomaly detection in Power BI.

Why use anomaly detection?

The reason we want to detect anomalies very much depends on the scenario. Let's go through some situations to help understand which data we can use when exploring the anomaly detection feature in Power BI.

Anomalies are per definition unexpected events, which means that they can come in many different forms. The main benefit of anomaly detection is that it can use *unsupervised ML algorithms*. Since it is unsupervised, we don't have to give the model examples to make sure it can actually detect anomalies in the future. Instead, it will find anomalies, even if they have never occurred before in the dataset we provide.

What, then, makes an anomaly? Sometimes, this is very straightforward. Think about outliers (as covered in *Chapter 2, Exploring Data in Power BI*)—data points that significantly differ from the rest of your data. For example, in the following screenshot, we have a box plot of the *Life Ladder* score per country in 2019, a measure for happiness per country taken from the *World Happiness* dataset (described in further detail in *Chapter 2, Exploring Data in Power BI*):

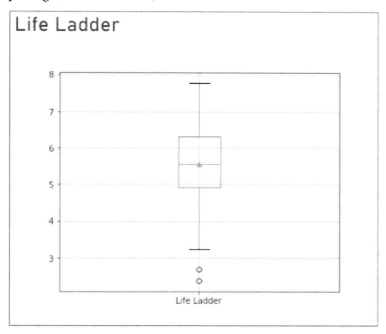

Figure 5.1 – Box plot of Life Ladder score

The box plot in *Figure 5.1* shows the distribution of the data points, and the two circles at the bottom of the plot represent outliers. In this case, the outliers are extremely low compared to the rest of the data and therefore unexpected and rare data points or, in other words, anomalies. As discussed in *Chapter 3, Data Preparation*, these outliers are not errors but actual realistic data. With these outliers, a simple box plot identified the anomalies for us, and we don't need to use any fancy anomaly detection algorithms.

Other times, however, it isn't so easy. Anomalies can be present in data even when no outliers show up when visualizing the data with a box plot. Next to that, anomalies can also occur over time, and especially when working with time-series data, you don't want to go through the effort of creating or refreshing a box plot every time you want to check for any anomalies.

For example, you can monitor the number of tweets that mention your brand or product name over time. You can expect the number of tweets to fluctuate every hour and every day. However, you are especially interested when that number suddenly increases or decreases in an unexpected manner. To identify anomalies, we can create a line chart, as illustrated in the following screenshot, which shows the number of tweets per day:

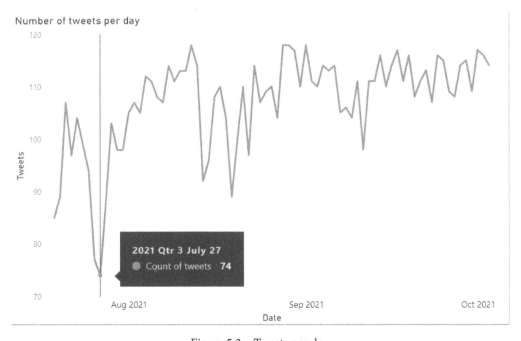

Figure 5.2 – Tweets per day

In this line chart, the number of tweets per day clearly fluctuates, which is expected. There are, however, a couple of dips that could be seen as anomalies. The first one occurs on July 27, 2021, with **74** tweets. Is it really an anomaly, though? Even though it is the first complete and clear dip since the data started to be collected, it may be normal for the number of tweets to fluctuate and go up and down regularly. A data point can differ from other data points but does not have to be an outlier. To more accurately judge whether a data point is an anomaly, it is better to use statistical methods than to simply draw conclusions from visualizations. Therefore, instead of visually investigating the data and deciding on anomalies, it may be better to have a model trained in recognizing trends and seasonality in order to clearly identify which data points are true anomalies.

In this section, we learned that anomalies—or unexpected data points—can be outliers. These outliers can be identified by creating a box plot of the data or by creating a line chart for time-series data to visually determine an anomaly. However, anomalies may be difficult to identify based on simply looking at any type of chart, which is why we want to use **unsupervised ML (UML)** algorithms to detect anomalies in our datasets. To use the out-of-the-box feature in Power BI for quick anomaly detection, let's go over what we need to have in our dataset.

Data requirements for anomaly detection

As mentioned briefly earlier, anomalies can come in many different shapes and forms. The type of anomaly you want to detect and the algorithm you use to detect anomalies depend on the data you have. Since Power BI's out-of-the-box anomaly detection feature uses a specific algorithm, we also need to make sure the data we put in meets the requirements, which is what we'll discuss in this section.

Requiring time-series data

One of the most common applications of anomaly detection is when monitoring a specific metric such as the temperature of an engine or the number of tweets mentioning your product. In such cases, the data is collected over time and therefore known as time-series data. To detect an unexpected measurement over time, we use **time-series anomaly detection**. And as you may expect, this is the type of anomaly detection Power BI currently supports. Since the out-of-the-box anomaly detection feature in Power BI is based on time-series data, there is one very important requirement of your data: it needs to include a `date` field.

Whichever metric you want to monitor, it has to be collected over time and your query needs to include a *field* that is of `date` type. The interval of time over which you collect data doesn't matter; you can have the number of tweets per minute or the temperature of an engine per hour. As long as there is a *DateTime value* or a *valid Date hierarchy value*, the anomaly detection feature in Power BI can be enabled. If you are missing such a field, the anomaly detection option will simply be grayed out. Having time-series data is the most essential requirement if you want to detect anomalies with Power BI.

Including enough data points

Another requirement is that you need enough data points for the anomaly detection feature to work. Remember that Power BI uses a UML method. Using UML means that there is no labeled data—there are no examples from which the model can learn.

The only way the model can identify anomalies is by understanding the data and its trends and seasonality (these concepts are explored in *Chapter 4, Forecasting Time-Series Data*). Data can fluctuate, and your metric can steadily increase or decrease over time. To understand what actual unexpected data points are, the model needs enough data points to analyze. At a bare minimum, you need *at least four data points* in order to do anomaly detection in Power BI. However, as is the case with many ML applications, the more data, the better.

Including other attributes

One last requirement to make use of anomaly detection in Power BI is optional. Next to detecting anomalies, we can also tell Power BI to try to explain what has caused a specific anomaly. If you want to make use of this extra feature, you need to *include other attributes in your data*. What do we mean by attributes? Think about the characteristics of the data measurements you include in your dataset. We'll explore what these attributes can be, but let's first try to understand what these attributes are not.

What do we not mean by attributes? In the previous chapter (*Chapter 4, Forecasting Time-Series Data*), we explored the number of tourists in the Netherlands. The number of tourists per month is shown in a line chart in the following screenshot:

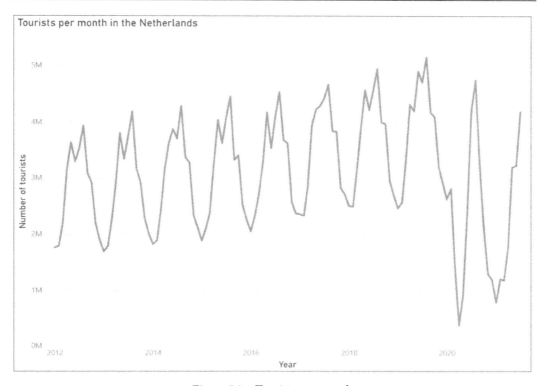

Figure 5.3 – Tourists per month

The number of tourists in the Netherlands increases steadily over time, though there is a natural seasonality per year. April 2020 shows an unexpected decrease in the number of tourists and could very well be an anomaly (we'll find out later what Power BI thinks). In this case, based on some context, we understand this is because of an external event: the COVID-19 pandemic.

The ability to draw a conclusion that the pandemic has caused a decrease in tourism is something very human. Including external data that we didn't include in the original dataset to explain data changes is not something AI can do. AI will only know what we feed into it. It is important to realize these limitations to ensure you know which data to include and what kinds of insights to expect.

So, what do we mean by *attributes*? We mean characteristics of the data measurements that have always been present in your data. So, in the case of the number of tourists in the Netherlands, this could be information such as whether the tourists live in the Netherlands or abroad, or whether they stayed at a hotel or camping site. These kinds of attributes can be included to help explain why there is an anomaly. In the case of the decrease in April 2020 (which we as humans know is because of the pandemic), it could be that the anomaly is caused by a decrease in tourists coming from abroad. Meanwhile, the number of tourists living in the Netherlands and going on vacation in the Netherlands has probably stayed the same or maybe even increased as borders closed and people chose domestic travel over international travel.

In the following screenshot, an example of a time-series dataset on tourism in the Netherlands is shown with attributes:

	Period	ABC Accommodation type	ABC Travel	123 Tourists
1	1/1/2012	Group accommodations	International	2000
2	1/1/2012	Hotel	International	563000
3	1/1/2012	Camping	Domestic	7000
4	1/1/2012	Hotel	Domestic	735000
5	1/1/2012	Group accommodations	Domestic	46000
6	1/1/2012	Holiday park	International	77000
7	1/1/2012	Holiday park	Domestic	328000
8	1/1/2012	Camping	International	5000
9	2/1/2012	Hotel	Domestic	749000
10	2/1/2012	Camping	International	4000

Figure 5.4 – Dataset on tourism in the Netherlands

The dataset in *Figure 5.4* has the following four fields:

- `Period`: The required date field.

- `Accommodation type`: The options are `Group accommodations`, `Hotel`, `Camping`, or `Holiday park`.

- `Travel`: This can be either `Domestic` (tourist lives in the Netherlands) or `International` (tourist lives abroad).

- `Tourists`: The number of tourists as a *whole number*. This is the metric we want to monitor.

In this dataset, `Accommodation type` and `Travel` will not initially be used to detect any anomalies but can be used once an anomaly has been detected to explain the difference with the rest of the data points. We will see this in action at the end of this chapter when exploring how to use anomaly detection in Power BI.

In this section, we learned that there are two important data requirements when you want to use the out-of-the-box anomaly detection feature in Power BI: having time-series data and enough data points. Optionally, you can include other attributes in your dataset if you want to be able to explain any anomalies based on some extra characteristics. Next, we will go over the algorithm used to detect anomalies in Power BI, after which we will try it out over the tourism dataset.

Understanding the logic behind anomaly detection

The algorithm behind the anomaly detection feature in Power BI has been developed by Microsoft and is designed for real-time time-series data in many different applications. The purpose of this section is not to go over all technical details but instead to keep it pragmatic. In this section, we will explore the key considerations of this algorithm that help us understand how to use the feature in Power BI appropriately. For more in-depth information, you can read the paper *Time-Series Anomaly Detection Service at Microsoft* by *Ren et al.*, 2019.

The algorithms behind Microsoft's anomaly detection feature

First of all, the algorithm created by Microsoft is a combination of two methods and is referred to as **SR-CNN**. SR stands for **Spectral Residual** and CNN stands for **Convolutional Neural Network**. Both of these **deep learning** (**DL**) methods are most often used for analyzing images, and it's the combination of these two techniques that make anomaly detection in real time possible on unlabeled data. The SR method essentially detects anomalies and gives them as input to the CNN method to further train the model to learn how to better detect anomalies. This way, you as a user don't have to provide a training dataset with anomalies, and the model is able to return anomalies quickly to you.

The anomaly detection feature in Power BI can only be enabled when plotting the metric of interest in a line chart with a `date` field on the x axis. What Power BI will then do is analyze the line chart to find any anomalies. Most likely, you will want to expand down to the lowest level in the date hierarchy to include as many details as possible. The model will only take the data it sees in the chart, not the data that exists in the table. So, if you don't include data in the chart, and instead opt for a higher level that aggregates the data, you may get different results than expected.

For example, if you plot the number of tweets per day, a complete day can be marked as an anomaly. If you want anomalies to be detected on a per-minute basis, you need to plot the number of tweets per minute instead.

Another thing to note is that DL methods (as described in *Chapter 1, Introducing AI in Power BI*) are used, and DL methods are known to always benefit from more data. So, the more data points you include, the better the model will be in detecting anomalies.

No need to label your data

There are other algorithms to use for anomaly detection as well. Microsoft has chosen to use a method that uses **unsupervised learning** (**UL**) (as described in *Chapter 1, Introducing AI in Power BI*). If instead, the anomaly detection is based on **supervised learning** (**SL**), you would have to label all known anomalies to make it possible for the model to identify new ones. In reality, this can be hard as some anomalies have never happened before or are too subtle, and in some cases, there may simply be too many anomalies to label. For this reason, the anomaly detection feature in Power BI uses UL, meaning you don't have to worry about labeling your data.

Fast and powerful analysis

One of the key reasons for using an out-of-the-box AI feature such as anomaly detection in Power BI is the time saved compared to creating and consuming a model you train yourself. The SR-CNN algorithm created for anomaly detection in Power BI has one key characteristic: it is fast. To make sure Power BI users such as data analysts experience minimal latency when enabling anomaly detection, the SR component plays a large role. SR is an algorithm optimized for finding visual saliency. Saliency is the noting of something that we humans deem important.

In other words, the algorithm looks at the data in the line chart and will focus on salient characteristics: changes in the plot that humans would consider interesting. It will then analyze whether that data is indeed different than expected. By using this approach, anomaly detection in Power BI is very fast and can quickly show results, even on large amounts of data. After all, it has been designed for real-time monitoring, during which a dataset can quickly grow to become very large.

In this section, we learned that Microsoft has developed its own algorithm to detect anomalies in services such as Power BI. It is designed to quickly analyze time-series data to detect anomalies by visually inspecting the data plotted in a line chart. Because of the DL methods used, there is no need to label your data, and the more data points you include, the better the model will be to detect anomalies. Now that we know how powerful anomaly detection in Power BI is, let's use it and explore the results.

Using anomaly detection in Power BI

Now that we have time-series data ready to go and understand what the anomaly detection feature in Power BI does, it's time to try it out.

Importing the sample dataset into Power BI

To follow along with the steps to enable anomaly detection in Power BI, make sure you have imported the tourism dataset. You can download the dataset from GitHub via the following link: `https://raw.githubusercontent.com/PacktPublishing/Artificial-Intelligence-with-Power-BI/main/Chapter05/tourism-details.csv`.

Save the downloaded dataset on your local system and name it `tourism-details.csv`. Import the dataset into Power BI and leave the query name as `tourism-details`, as it will be referenced by that name. The query will consist of four fields: `Period` (categorized as date), `Accommodation type` (categorized as text), `Travel` (categorized as text), and `Tourists` (categorized as whole number).

Enabling anomaly detection in Power BI

Once you have the data in Power BI, you can follow these steps to enable anomaly detection. The steps assume you have **Auto date/time** enabled in Power BI:

1. Create a line chart with **Period** on the axis and **Tourists** as the value.

2. You want to include as many data points as you can. Therefore, select **Expand all down one level in the hierarchy**, to the lowest level in the **Period** hierarchy. You'll get the following result:

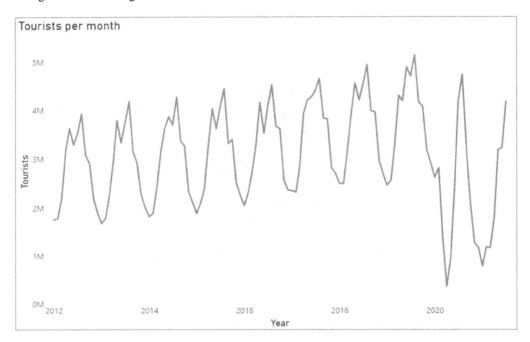

Figure 5.5 – Line chart of tourists per month

3. There are two ways to enable anomaly detection, as follows:

    A. In the top ribbon, select the **Data/Drill** tab. Then, select **Find anomalies**.

    B. In the **Visualizations** pane, select the **Analytics** tab. Then, expand **Find anomalies** and select **+Add**.

    Note that in the **Analytics** tab in the **Visualizations** pane, you can edit the formatting of the anomalies under **Find anomalies**. You can edit the shape of the icons with which anomalies are marked in the line chart. You can edit the size of the icon as well as the color. You can also change the style, color, and transparency of the shaded area representing the expected range of values.

    Once enabled, a shaded area will be added to the line chart and any anomalies will be identified with an icon, as shown in the following screenshot:

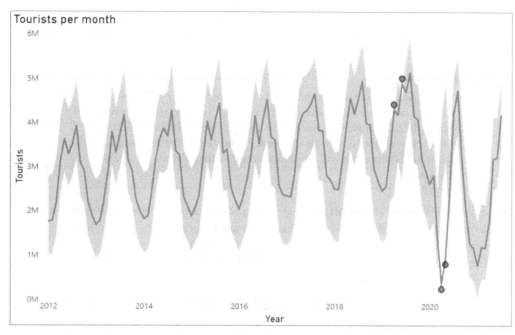

Figure 5.6 – Anomalies for tourists per month

As we can see from *Figure 5.6*, there are four anomalies detected in the tourism dataset. The latter two anomalies occurring at the beginning of 2020 were expected because of the COVID-19 pandemic, as discussed earlier in this chapter. The first two anomalies, however, are actually more unexpected. In April and June 2019, there were more tourists than the model expected.

The anomalies in 2020 are an example of anomalies that we already expected as a result of visually inspecting the data. When creating the line chart, we could already see a clear trend and seasonality, and the dip at the beginning of 2020 did not fit with that trend. We could have already identified those anomalies. The anomalies in 2019, however, are interesting because we didn't expect those based on visual inspection of the line chart. This shows that anomalies can be obvious but also can be very subtle.

Now, let's try to see if we can dive a little deeper and understand more on why these anomalies have occurred. As explained earlier, we can't expect Power BI to know that COVID-19 affected tourism since we didn't include that in the dataset (and didn't expect the need to include it until the anomaly actually occurred). Other attributes that we included were `Accommodation type` and `Travel`.

By default, Power BI will try to see if there is any relevant data that it can use to explain the anomalies.

> **Important Note**
>
> When using features such as anomaly detection, you'll notice that Power BI will try to find correlations between your variables. The correlations presented by Power BI do not always make sense. To verify the correlations, someone with domain knowledge of the data is needed. Only by understanding what the variables or fields represent will you know whether to conclude a correlation is indeed valuable.

To force Power BI to look at certain correlations (and not look at other fields), it is a best practice to include the fields you want Power BI to use to explain the anomalies. In our case, we want Power BI to look at how `Accommodation type` and `Travel` may have affected the number of tourists. If we don't specify both these fields, it may be that Power BI only looks at one of the two.

To add fields you want Power BI to use to explain anomalies, follow these steps:

1. Go to the **Visualizations** pane, then to the **Analytics** tab, and then to **Find anomalies**.

2. Drag the `Accommodation type` and `Travel` fields to the **Explain by** box.

3. Select **Apply**.

4. Select the first anomaly occurring in **April 2019**.

   The **Anomalies** pane will open on the left of the **Visualizations** pane. An example of the pane is shown in the following screenshot, where the first anomaly is selected and described below the chart:

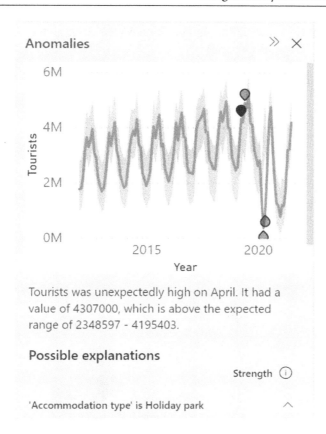

Figure 5.7 – Anomalies pane

The **Anomalies** pane gives us extra insights into why anomalies occurred. First of all, we see a brief summary of the data point we selected. In this case, it says that the number of tourists was unexpectedly high in April 2019 compared to what it would have predicted. Below that summary are possible explanations for that specific anomaly: why were there more tourists in April 2019? One explanation seems to be that more people were staying in holiday parks, which increased the total.

This may not give a full picture of the cause of the anomaly, but it does give you an idea of where to look for potential causes. Maybe holiday parks advertised more that year or simply became more popular. More domain knowledge could be needed to draw proper conclusions on this.

As a comparison, let's look at the anomaly in May 2020.

5. Select the anomaly in **May 2020** (can be done on the original line chart or on the line chart in the **Anomalies** pane).

The **Anomalies** pane will now show a summary of the anomaly in May 2020, stating the number of tourists for that month was unexpectedly low, as shown in the following screenshot:

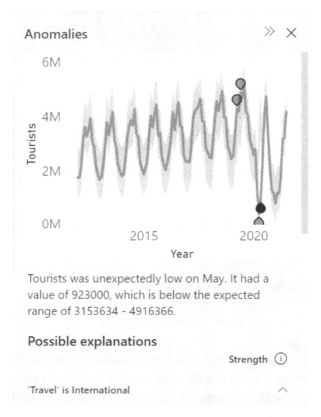

Figure 5.8 – Anomaly summary for May 2020

Underneath the summary in *Figure 5.8*, we can see already one possible explanation: **'Travel' is International**. In the following screenshot, we can see the accompanying chart:

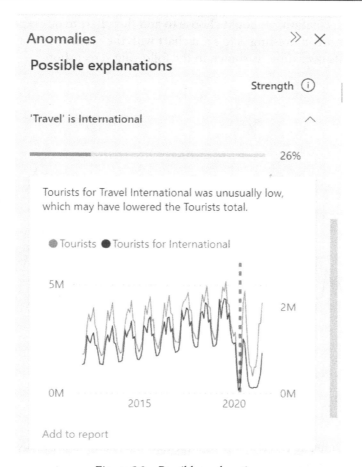

Figure 5.9 – Possible explanation

As seen in *Figure 5.9*, the number of tourists being unusually low in May 2020 is possibly related to the lower number of international tourists that month. Based on the findings of the anomaly detection feature, Power BI thus plots the international tourists next to the total tourists to show us what may be causing the anomaly for the month of May.

In other words, although the number of tourists living in the Netherlands and going on vacation in the Netherlands was as expected, the number of tourists living abroad visiting the Netherlands was lower than expected, causing a significant dip in the total number of tourists.

If this is valuable information, we could choose to add this chart to our report. In this case, what may be more interesting is to see a chart with the number of both international and domestic tourists over time, as shown in the following screenshot:

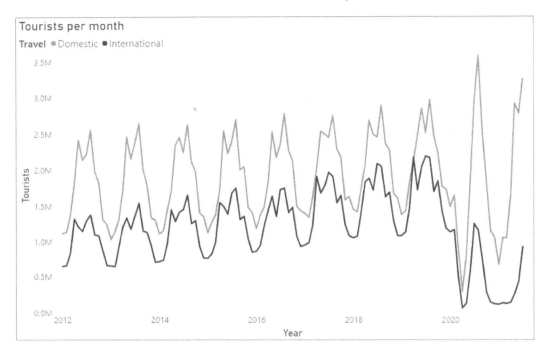

Figure 5.10 – Domestic and international tourists

Figure 5.10 shows domestic travel as the top line and international travel as the bottom line. Before 2020, we notice a repeating pattern or seasonality, which is then disrupted at the beginning of 2020. In April 2020, almost no one was traveling. After that dip, it seems domestic travel picked up more quickly than international travel, which accounts for Power BI's perception of the anomaly being caused primarily because of fewer international tourists.

Again, to make proper conclusions, we need a human to understand how the attributes are related to the metric we are monitoring and to see if other factors may have influenced those attributes. For example, in this case, the COVID-19 pandemic could indeed have resulted in fewer tourists coming from abroad. Nevertheless, Power BI can very quickly give us valuable insights into any anomalies that may occur in the data, even if we didn't see these ourselves when looking at outliers in box plots or obvious spikes or dips in line charts.

Summary

Power BI can help us to get quick insights into our data. One powerful out-of-the-box feature is anomaly detection. To use anomaly detection, we now know we need to have time-series data with enough data points and—optionally—some other attributes we want Power BI to use to explain any anomalies it may find. We learned that by plotting the metric we want to monitor in a line chart with a date field on the x axis, we can enable anomaly detection to find any anomalies. Some anomalies may be obvious, and some may be very subtle. For whichever anomaly Power BI finds, we can explore any possible explanations based on the attributes we included in our dataset to get insights into why these anomalies have occurred. Remember that domain knowledge is still needed to validate these insights and correctly interpret them. In the next chapter, we will look at another out-of-the-box feature: the **question-and-answer (Q&A)** visual. This visual not only helps you but actually allows readers of shared reports to get quick insights into the data based on natural language.

Further reading

To further understand the concepts of this chapter, you can refer to the following links:

- Official Microsoft documentation on anomaly detection in Power BI: `https://docs.microsoft.com/en-us/power-bi/visuals/power-bi-visualization-anomaly-detection`

- *Overview of SR-CNN algorithm in Azure Anomaly Detector*: `https://techcommunity.microsoft.com/t5/ai-customer-engineering-team/overview-of-sr-cnn-algorithm-in-azure-anomaly-detector/ba-p/982798`

6
Using Natural Language to Explore Data with the Q&A Visual

Instead of being restricted to the insights of the data shown in the visuals created by data analysts, the Q&A visual allows users to ask questions about the data in a natural way and get answers through visuals.

Suppose you are a data analyst and you have created reports to share with a large number of people. Not everyone wants to see the same visuals or insights of the data. By adding a Q&A visual to your report, you allow users to ask questions about the data and get insights in the form of visuals, all without the users needing to know how to create the visual to get the necessary insight.

In this chapter, we're going to cover the following main topics:

- Understanding natural language processing
- Creating a Q&A visual in Power BI
- Optimizing your Q&A visual

After we've covered these topics, you'll be able to add a Q&A visual to your Power BI report. To make sure your users get sensible answers and the insights they are looking for, you will learn how to configure the visual.

Technical requirements

There are two things you need to walk through the examples provided in this chapter:

- Power BI Desktop:

  As this book revolves around Power BI, we expect you to have Power BI Desktop installed on your system. You can install Power BI from the Microsoft Store or find more advanced downloading options here: `https://www.microsoft.com/en-us/download/details.aspx?id=58494`.

- Sample dataset:

  A sample dataset is used in the examples provided in this chapter. The data is the financial sample dataset that you can download directly in Power BI Desktop. Simply open Power BI Desktop and once you see a blank canvas, select **Try a sample dataset**. Learn more about the dataset and how to download it here: `https://docs.microsoft.com/en-us/power-bi/create-reports/desktop-excel-stunning-report`.

Understanding natural language processing

Before we can explore how to create the Q&A visual in Power BI, let's first try to understand what it actually does and what its purpose is. For that, we need to discuss the term **natural language**. While on the subject of natural language, we'll learn what the most important data requirements are for natural language Q&A on your data to work.

Using natural language in programs

The number of tools that integrate with natural language models is increasing. What is this natural language that is advertised everywhere? Let's explore the idea and purpose of natural language.

The basic premise of natural language is that we want to make it more intuitive for people to work with programs. Instead of us having to learn how to explore data with SQL or how to create a line chart with Power BI, the idea is that we can say what we want and the program will do the work for us.

In other words, we want programs to work with us, as if they are our colleagues. Let's assume you are a data analyst. Your colleagues may ask you to create a Power BI report for them that includes specific insights. They may say, "I want to see which month and year we were the most successful in selling our products." You then translate that question into a chart that could represent that information, based on the data that you have in your database.

Translating a question such as that into a technical process to create a chart is what natural language models do. Whenever we talk about using natural language in applications, we mean that we want users to use the language they would normally use when talking to another person. Using natural language to find information is becoming more and more valuable for businesses, as natural language is something everyone knows, but using a tool is something people must invest in to learn.

But why are we talking about natural language? To answer that, we can use another great example. Imagine you want to find out what the biggest country in the world is. You can go online, use any search engine, and type in `what is the biggest country in the world?`. The answer will be given to you almost instantly. If you had used that same search engine 10 years ago, you would probably have had to rephrase your question. You would have typed something such as `big country world` or `list world countries size`. You knew the search engine was using specific keywords to find information, and you would try your best to make it as easy as possible for the search engine to understand what you were looking for.

So, we can use specific keywords, selected carefully for the search engine we want to use. Or we can simply ask a question like you would ask a question to another human person. The latter is more natural to us and, therefore, called natural language. Natural language is what we want to aim for as it is the most intuitive method for people to work with.

Natural language is currently used in chatbot conversations, to search for relevant information in large amounts of text, but also for exploring data through Q&As. We'll discuss how we can use natural language for data exploration next.

Understanding natural language for data exploration

Exploring data can be a time-consuming task. A table of data can hold a lot of information and there can be many different perspectives you can take to extract insights from the data.

As a data analyst, you are trained to understand how to explore data using Excel, Power BI, and maybe even SQL queries to extract insights.

To help you in your work, you can use natural language for data exploration. This subfield of natural language models is focused on translating text into queries performed on a database and is often referred to as **database question answering** (**DBQA**).

For example, your colleague may ask you to show in which month we were most successful in selling our products. To get the answer, you can perform a SQL query where you sort the month column by the values in the profit column, as you can see in *Figure 6.1*:

Figure 6.1 – Converting a question to a SQL query

The purpose of DBQA is to translate a question asked in natural language to a query that can perform data exploration on structured data. Since we are essentially translating, the logic behind DBQA is largely based on the work done in translating text from one language to another. At the time of writing, the Q&A visual is likely to use **Structured and Sequential Context Representation** (**ScoRe**), a language model designed by Microsoft Research. To read more about the technical details of this model, you can read the research paper here: https://www.microsoft.com/en-us/research/publication/score-pre-training-for-context-representation-in-conversational-semantic-parsing/. Here, we will focus on the general idea of the language model and what we need to understand to use the Q&A visual correctly.

The SCoRe model is designed to translate natural language into a programming language such as SQL in order to query a database. One important aspect of this model is that it is pretrained and it generalizes well over different databases. These language models are deep learning models, which means that they require a lot of time and compute to train. Having a pretrained model will therefore save you time and money as a user.

Whenever you use the Q&A visual, the model will look at your dataset and define key-value pairs. The keys are the column names in your dataset and the values are the rows in your dataset. Based on the computations we can do on data, such as aggregations or filters, the model starts to imagine the kind of queries you may want to do.

The model will combine the key-value pairs and the possible computations, and will suggest questions that can be asked. As we will see later when we create a Q&A visual in Power BI, these suggestions pop up as soon as you add this visual to your report.

The most important thing the model is trying to do is to match a question to the key-value pairs in your data and the computations that can be done on that data. The more similar a word in your question is to a key or value in your dataset, the easier it is for a model to match it. The matching is called **semantic alignment**.

There are two parts to semantic alignment:

- **Phrasing the question**: The Q&A visual will correct the spelling of words used in a question. It will also autocomplete and autosuggest words and questions. It will give feedback to the user if words are not recognized based on the dataset and its capabilities, helping the user to phrase the question correctly. Phrasing the question correctly makes it easier for the model to do semantic alignment.

- **Finding the answer**: To extract answers from the dataset, the model needs to understand the data. It is easier for the model to match a question to an answer if the same words are used. To help the model to find the most relevant answer, the data needs to be optimized for natural language querying.

To help with phrasing the question, we can optimize the Q&A visual, as we will discuss later in this chapter. To help with finding the answer, we can prepare the data to help the model understand what information is in our dataset. We will discuss how to prepare data for the Q&A visual next.

Preparing data for natural language models

Whenever we use natural language to query our data, the model will try to semantically match words in our question to words in our data in order to form an answer. The better prepared your dataset is, the easier it is for the model to find the relevant answer to your question. Let's explore how we can optimize our data for when we want to use the Q&A visual in Power BI.

Focusing on the core functionality of the Q&A visual, semantic matching, you want to do the following:

- **Make table and column names more humanly intuitive**: Avoid abbreviations and combined words. So, instead of `sls` or `salesamount`, use `sales` or `sales amount`.

- **Select the correct data type for each column**: Check that the correct data type is selected, such as `date`, `number`, or `text`.

- **Select the appropriate data category for each column**: Add additional semantic information to each column by selecting the data category, such as `city`, `country`, or `zip code`.

- **Add missing relationships**: If you're working with multiple tables, define relationships between the tables using primary and foreign keys. The model will only use existing relationships.

For more guidelines on preparing data for Q&A in Power BI, check out the following link: `https://docs.microsoft.com/power-bi/natural-language/q-and-a-best-practices#add-missing-relationships`.

Power BI's Q&A visual uses a language model to explore data through natural language. By asking a question, the model will formulate an answer based on the dataset in Power BI. When preparing the dataset, it is easier for the model if intuitive table and column names are used, the correct data type and category are selected, and all necessary relationships are in place.

Creating a Q&A visual in Power BI

Now that we understand the language model behind Power BI's Q&A visual, let's add the visual to a report and explore how users interact with it.

Adding a Q&A visual

For this example, we will use the financial sample in Power BI:

1. Open a new, blank Power BI report.

2. Download the dataset by selecting **Try a sample dataset**. We'll use the financial sample dataset that includes a fictional company's sales data. The data includes the sales for different products, per month, for the years 2013-2014. For more information on this dataset, check out the following link: `https://docs.microsoft.com/power-bi/create-reports/desktop-excel-stunning-report`.

3. Select **Load data**.

4. Select **Financials** and **Load data**.

    Once the dataset is imported into Power BI, you should see a query in the **Fields** pane named `financials`.

    From the **Visualizations** pane, select **Q&A**.

    A Q&A visual is added to the report, similar to the one shown in *Figure 6.2*:

Figure 6.2 – Q&A visual

The Q&A visual will show suggested questions, based, among other things, on the column names of your dataset. In this case, the `COGS`, `Product`, and `Date` fields of the `financials` table have contributed to the suggested questions.

Based on possible questions using natural language, the model already did some semantic matching. For example, instead of `Date`, the model suggests `time`. A user is more likely to ask what happens to a variable over time than what happens to it over a date, the former being more in line with what we consider to be natural language.

Using the Q&A visual

To reenact the user's experience when they get access to the report, we can interact with the Q&A visual to test it.

Users will be able to select one of the suggested questions or type their questions into the search bar. Let's try it out ourselves with the Q&A visual we created in the previous section:

1. Select the suggested **total cog over time** question, or type it into the search bar.

    The result that the Q&A visual gives may vary. An example output is a line chart, as shown in *Figure 6.3*:

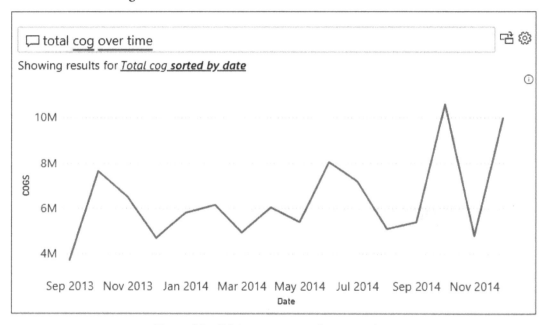

Figure 6.3 – Q&A answer to total cog over time

The line chart plots the COGS field as values on the *y* axis and the Date field as the *x* axis. Below the search bar, it even shows a glimpse of the translation of the question into the programmatic query that is performed on the dataset: **Total cog sorted by date**. This shows us that the model semantically matched time with date.

Let's try another question.

In the search bar, replace the current question with which month and year had the most profit.

The result may vary, but as an example, we may get another line chart, as shown in *Figure 6.4*:

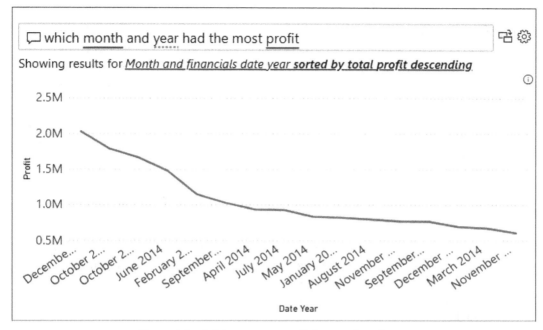

Figure 6.4 – Q&A answer to profit by month and year

Again, we can see which fields have been used, based on the sentence after **Showing results for**, and the information in the line chart. This time, `Profit` is plotted against `Date Year`, which is sorted by profit per month.

When looking at the question in the search box, you may also notice that words can be underlined in three different ways:

- **A single underline**: Means the model understands the word.

- **A single dashed underline**: Means the model understands the word but is not too sure because there may be another way to interpret the word. If you select this word, you can see alternative interpretations.

- **A double underline**: Means the model does not understand the word. If you select this word, you can select to define the word to add it to the model. This is discussed in the *Optimizing your Q&A visual* section in this chapter.

When asking a question, it helps to understand a little bit about how the question is interpreted to form an answer. The Q&A visual in Power BI can create some calculated views of the data based on words it picks up, such as *the top five selling products*.

To educate end users and help them formulate their questions so that they get the results they need, you can share some tips with them. For example, you can include the tips listed by Microsoft, which you can find here: `https://docs.microsoft.com/en-us/power-bi/consumer/end-user-q-and-a-tips#words-and-terminology-that-qa-recognizes`.

In this section, we have seen how we can quickly and easily create a Q&A visual to add to a Power BI report. We learned how we can use it and how to interpret its results. Now, let's have a look at how to optimize a Q&A visual to gradually improve it over time.

Optimizing your Q&A visual

We have learned that the language model used by the Q&A visual is pretrained to semantically match questions asked by a user to answers extracted from the dataset in Power BI. The model is trained to generalize well over different datasets, which means it can make mistakes. To improve the user's experience, the Q&A visual can be optimized when opening the report with Power BI Desktop. Let's explore how we can improve the language model.

Exploring the Q&A setup

Any improvement or optimization we want to do on the language model used by the Q&A visual can be done through the Q&A setup in Power BI Desktop. To open this setup, we'll use the following steps:

1. Make sure a Q&A visual is added to your report in Power BI Desktop.
2. Select the settings icon on the top right of the Q&A visual (highlighted with an arrow in the following figure), as shown in *Figure 6.5*:

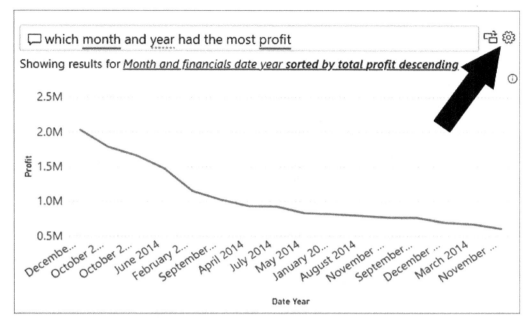

Figure 6.5 – Settings icon in Q&A visual

After you select the settings icon in the Q&A visual, as shown in *Figure 6.5*, a new pop-up window will appear with the Q&A setup. In *Figure 6.6*, the **Getting started** tab is shown:

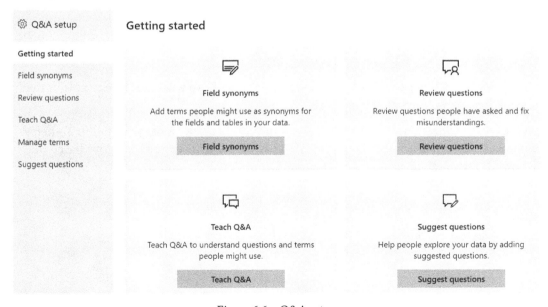

Figure 6.6 – Q&A setup

Opening the Q&A setup shows an overview of the things we can do to improve the language model:

- **Field synonyms**: For each table and column name, you can add synonyms or alternative terms with which users can refer to the same field. The Q&A setup will give you suggested terms to add as synonyms, generated by the language model. If you want a field not to be used to form an answer to a user's question, you can exclude it from the Q&A visual.

- **Review questions**: Shows all questions asked through the visual in the last 28 days. You can see the questions from users, including an underline to indicate whether the model understood it and whether the user requested a fix because of misunderstood words.

- **Teach Q&A**: Test possible questions and improve the result immediately by adding field synonyms and defining terms.

- **Manage terms**: After defining terms in the **Teach Q&A** tab, you can revise them in the **Manage terms** tab.

- **Suggest questions**: Add suggested questions that the user will see underneath the search bar when they interact with your report. This will guide users on the questions they can ask or serve as a quick link to frequently asked questions.

As we can see from the Q&A setup, there are many ways that we can improve the language model that finds answers for our questions based on our dataset. Let's explore how we can use these options by using the financials sample dataset in the next section.

Improving the Q&A experience

To learn how to improve the Q&A experience in Power BI, we will take the example of the financials dataset we imported into Power BI.

Imagine a user asking a question such as *how much profit did we have per month?*. To help them phrase their question correctly, the Q&A visual will autosuggest questions based on the fields it finds in the dataset, as you can see in *Figure 6.7*:

Figure 6.7 – Autosuggestions

In this figure, we see that the visual is suggesting we want to see `profit` per `month` name or per `month number`. The latter is something we as data analysts may use for ordering the months correctly in visuals but is not a field you want users to see or use. We can exclude this field from the Q&A visual so that users will not see this suggestion anymore, nor will `month number` be used to visualize any answer.

We can exclude a field in the Q&A setup as follows:

1. In the **Q&A setup** window, select the **Field synonyms** tab.

2. Expand the `financials` table. You should see all columns in the `financials` table, as shown in *Figure 6.8*:

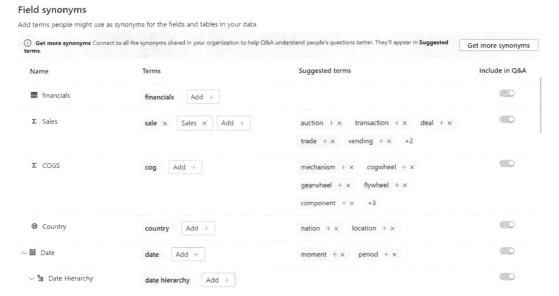

Figure 6.8 – Field synonyms

The field synonyms show the names of tables and columns and the terms or keys that the language model semantically matches with the names. You can add more terms to make it easier for the model to match a question to an answer. Possible additional terms are shown under **Suggested terms**. Lastly, we see a toggle to include or exclude a field in the Q&A visual.

3. Scroll down to `Month Number` and select the toggle to exclude this field from the Q&A visual. The result should look as in *Figure 6.9*:

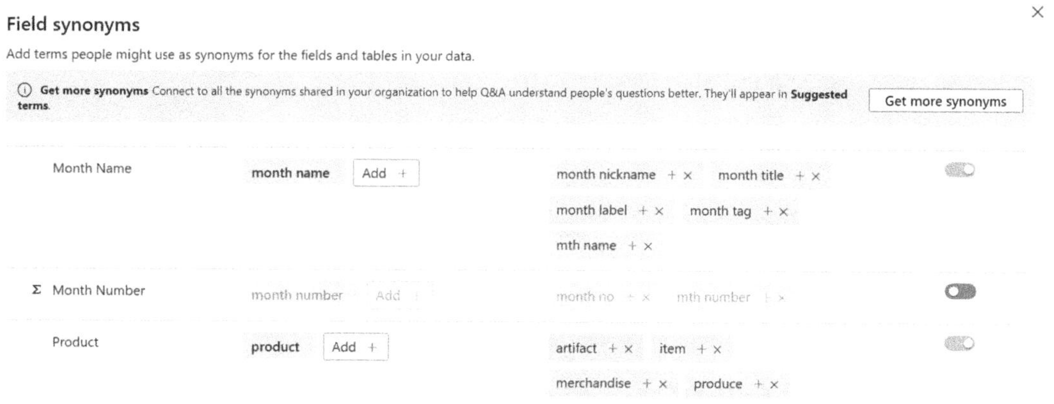

Field synonyms   ×

Add terms people might use as synonyms for the fields and tables in your data.

ⓘ **Get more synonyms** Connect to all the synonyms shared in your organization to help Q&A understand people's questions better. They'll appear in **Suggested terms**.   [Get more synonyms]

Month Name	**month name**	Add +	month nickname + ×	month title + ×	
			month label + ×	month tag + ×	
			mth name + ×		
Σ Month Number	month number	Add	month no + ×	mth number + ×	
Product	**product**	Add +	artifact + ×	item + ×	
			merchandise + ×	produce + ×	

Figure 6.9 – Month Number excluded

As shown in this figure, `Month Number` is grayed out as an indication that it still exists in the dataset but is not used by the Q&A visual.

You may have noticed that the model was able to produce a correct answer when asked the question *how much profit did we have per month?*. It is highly likely that users will formulate a question like this; instead of `month name`, a question will simply include `month`. To ensure this is always matched correctly, we can add the `month` synonym to `Month Name`.

The difference between `month` and `month name` is of course very small, and the language model is very adept at semantically matching these terms. A field that probably requires a synonym more is the `COGS` field. This abbreviation may be commonly used by your report's users, but we still want to make sure that if people use the full definition instead of the abbreviation, the Q&A visual finds the correct answer for them.

4. In the **Field synonyms** pane, scroll up to the COGS field. We can see the suggested terms given by the model, as shown in *Figure 6.10*:

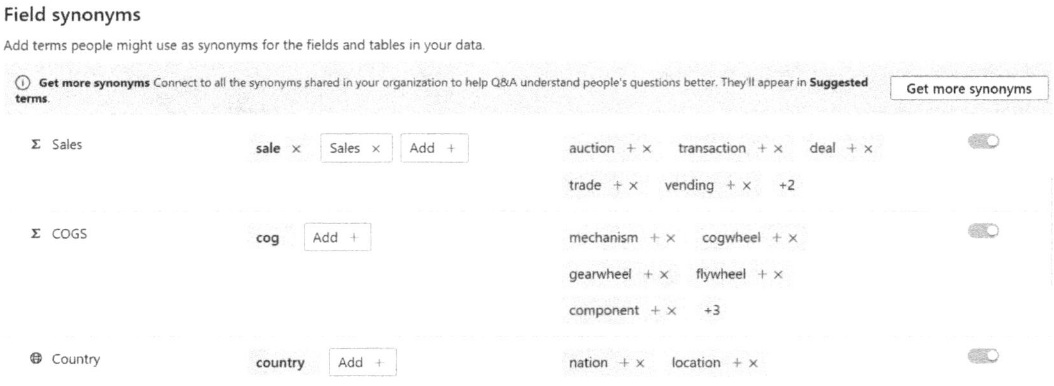

Field synonyms

Add terms people might use as synonyms for the fields and tables in your data.

Figure 6.10 – Suggested terms for COGS

Notice, in this figure, that suggested terms for COGS are terms such as mechanism, cogwheel, and gearwheel. This tells us that the model thinks COGS represents a cogwheel instead of **cost of goods sold**. Since the model will try to semantically match column names with words used in questions, we need to make sure the model understands what we mean by a word. Therefore, we need to add some synonyms here to improve the model.

5. In the **Terms** column for COGS, Select **Add +**.

6. Add synonyms to COGS, such as cost of goods sold, cost of goods, and cost. As you add synonyms, the list of suggested terms will change too, as shown in *Figure 6.11*:

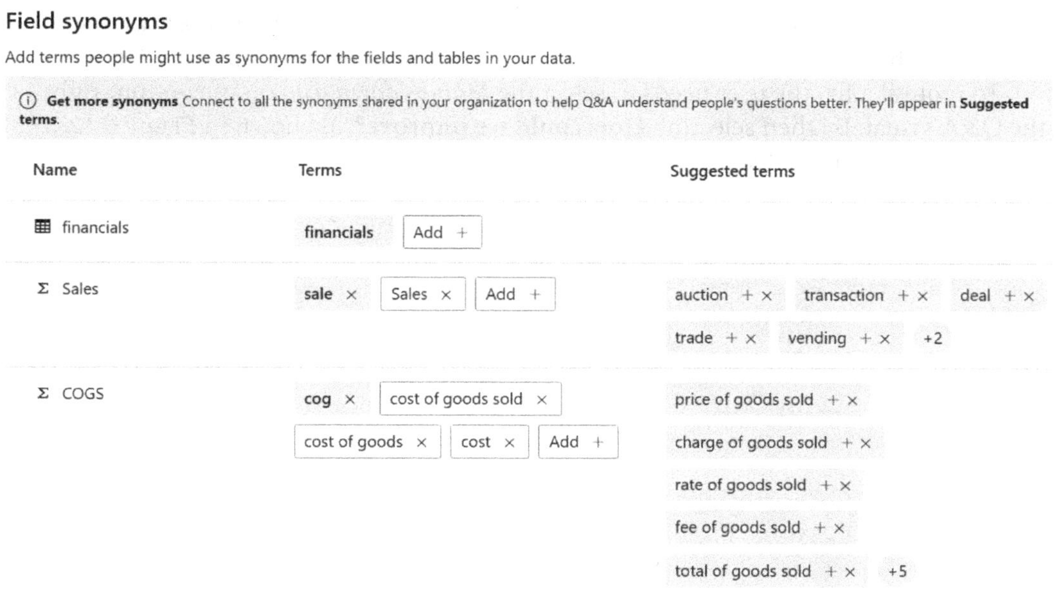

Field synonyms

Add terms people might use as synonyms for the fields and tables in your data.

ⓘ **Get more synonyms** Connect to all the synonyms shared in your organization to help Q&A understand people's questions better. They'll appear in **Suggested terms**.

Name	Terms	Suggested terms
⊞ financials	**financials** [Add +]	
Σ Sales	**sale** × [Sales ×] [Add +]	auction + × transaction + × deal + × trade + × vending + × +2
Σ COGS	**cog** × [cost of goods sold ×] [cost of goods ×] [cost ×] [Add +]	price of goods sold + × charge of goods sold + × rate of goods sold + × fee of goods sold + × total of goods sold + × +5

Figure 6.11 – Improved COGS

Now that we have added the synonyms, the list of suggested terms for COGS shows us that the language model better understands what we mean by COGS. Even if a user uses a different term than the ones we added as synonyms, the language model will be more likely to provide the correct answer based on the semantic meaning.

Using feedback to improve the model over time

The most important thing to remember with a language model such as the one used in the Q&A visual is that it will improve over time. Often a comparison is made between deep learning models and the way children learn. A child learns to speak through examples, practice, and feedback. The same goes for the model supporting the Q&A feature in Power BI. It will learn over time through examples, practice, and feedback.

Especially in the beginning, it is crucial to regularly review questions in the Q&A setup to ensure the model returns correct information to your users. Any model will make mistakes. It is how we handle those mistakes that will make it a positive or negative experience for the user.

Whenever a user asks a question that the model does not understand, the user can also *request a fix*, which will highlight a question in the **Review questions** tab of the Q&A setup. To request a fix, the user needs to select the *thumbs down icon* on the bottom right of the Q&A visual. By then selecting **How could we improve?**, as shown in *Figure 6.12*, a popup will appear to give feedback:

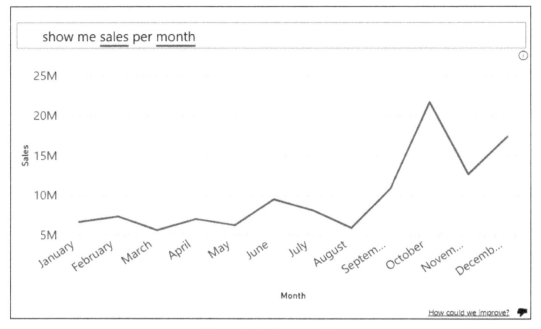

Figure 6.12 – Request a fix

The resulting popup, as shown in *Figure 6.13*, will ask the end user to formulate their feedback:

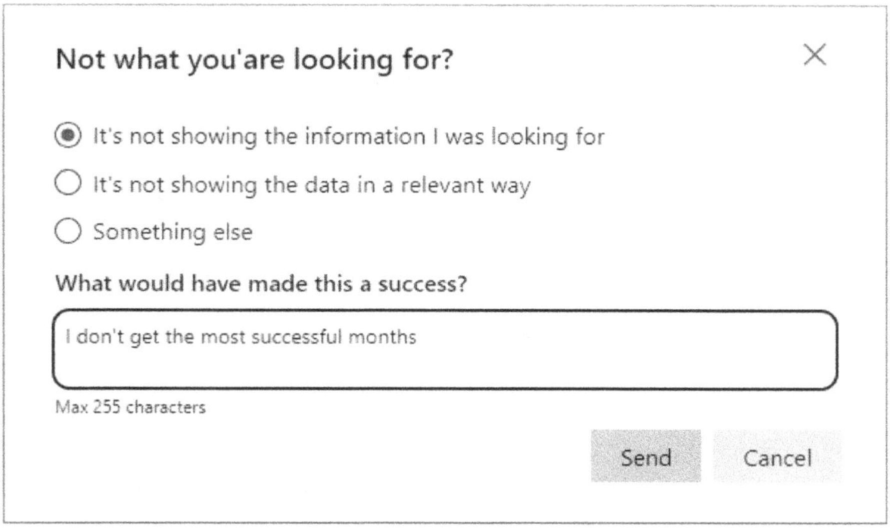

Figure 6.13 – Give feedback

Alternatively, as the creator of the report, you can add a text box to inform the user *who to email to request a fix*.

Improving the Q&A visual can be done both proactively, by adding synonyms, as well as reactively, by reviewing asked questions and requested fixes. Over time, the language model supporting the Q&A visual will improve.

Summary

In this chapter, we learned what natural language is and why we would want to use it. We learned how to use the Q&A visual in Power BI, which uses a language model that translates a question into a query that can be performed on a Power BI dataset to return an answer. The Q&A visual allows users to get insights through natural language questions. The language model used by the Q&A visual semantically matches words from the user's questions to the terms used in the dataset. We can help the model to find the correct answer by preparing the data and by adding synonyms to table and column names to facilitate semantic matching. By also reviewing asked questions and feedback from users, we can improve the model over time.

In the next chapter, we will learn about an Azure suite of services that we can integrate with Power BI to get access to more AI models that can enrich our datasets.

Further reading

- Microsoft Research blog on language-driven data exploration: `https://www.microsoft.com/en-us/research/blog/conversations-with-data-advancing-the-state-of-the-art-in-language-driven-data-exploration/`

- Research paper on the SCoRe model: `https://www.microsoft.com/en-us/research/publication/score-pre-training-for-context-representation-in-conversational-semantic-parsing/`

- Tips for asking questions in Power BI Q&A: `https://docs.microsoft.com/en-us/power-bi/consumer/end-user-q-and-a-tips#words-and-terminology-that-qa-recognizes`

- Best practices to optimize Q&A in Power BI: `https://docs.microsoft.com/en-us/power-bi/natural-language/q-and-a-best-practices#add-missing-relationships`

- Edit Q&A linguistic schema and add phrasings in Power BI: `https://docs.microsoft.com/en-us/power-bi/natural-language/q-and-a-tooling-advanced`

7
Using Cognitive Services

Azure provides **application programming interfaces (APIs)** with prebuilt **artificial intelligence (AI)** models to save time and money when trying to integrate AI into your solutions. In this chapter, we'll learn which models we can use and when.

Many organizations already see the value of **machine learning (ML)** models but do not have the resources to invest in them. Power BI users can easily and quickly make use of pretrained models with **Azure Cognitive Services**. Using these services, you can easily extract insights from unstructured data such as text and images.

When visualizing customer feedback, you can use Azure Cognitive Services to visualize the average sentiment of customer feedback over time. For example, imagine being able to not only track Twitter data on your company's brand but also find out whether people are talking about your brand positively or negatively.

To learn what exactly we can extract from text and images, we'll learn more about the following topics:

- Understanding Azure's Cognitive Services
- Using Cognitive Services for **language understanding** (**LU**)
- Using Cognitive Services for **Computer Vision** (**CV**)

After this chapter, you'll know what Azure's Cognitive Services are and how you can set up these services before integrating them with Power BI.

Technical requirements

To follow along with the examples in this chapter, you'll need the following:

- **An Azure subscription**

  The Cognitive Services is a service available in Azure. To create a resource, you need to have access to an Azure subscription. If you don't have one yet, you can sign up for a free subscription here: https://azure.microsoft.com/en-us/free/.

- **Code snippets and data**

  You can find the complete code files referred to in this chapter in the GitHub repository using the following link: https://github.com/PacktPublishing/Artificial-Intelligence-with-Power-BI/tree/main/Chapter07.

  All data used in this chapter (reviews and images) has been created by the author.

Understanding Azure's Cognitive Services

Throughout this book, we've been exploring how to use AI and integrate it with Power BI. The trade-off between building your own models from scratch and using prebuilt models is often mentioned. In later chapters, we'll find different ways to get more control over the data used to train the model and the way the model is trained. In this chapter, we'll introduce another approach to use out-of-the-box AI features: with Azure Cognitive Services.

Azure Cognitive Services is a suite of services where the main intention is *to reduce the time and expertise* needed to apply AI within your organization. Most of the prebuilt models listed under Cognitive Services are AI solutions to problems many organizations face.

For example, you may want to make your documents more accessible to people who are visually impaired and use programs that read the text out loud to them. If you include any images in your documents, the program may have difficulty understanding it. By adding **alternative text** (**alt text**), you make sure that the program can also describe the image to the user. You can add this alt text manually by having someone go through all images and give a description, but you can also use a prebuilt model to generate descriptions for all images. This application can be useful to many companies, and instead of each organization having to train its own model, it is beneficial for all to use a similar model as it is used for a similar purpose.

Many AI models make up the Cognitive Services. To use a model, we can make an API call, which basically means we send a request to an endpoint and get a response back. This is also why we sometimes refer to the Cognitive Services as a suite of APIs we can use to deploy different models. The types of models are divided into the following four categories:

- **Language**
- **Vision**
- **Speech**
- **Decision**

Here, we want to focus on AI models that can enrich our Power BI reports. Not all Cognitive Services will be of use to us in this scenario. We will focus on the Cognitive Services for *Language* and *Vision*. Feel free to explore more of what the other Cognitive Services can do by going through the following documentation: `https://docs.`
`microsoft.com/en-us/azure/cognitive-services/what-are-`
`cognitive-services`.

Now that we have a general idea of what the Cognitive Services can do for us and what their purpose is, let's create a resource so that we can subsequently see some of the features in action.

Creating a Cognitive Services resource

To use any of the features of the Cognitive Services suite, you have to create a Cognitive Services resource in Azure. There are multiple ways of creating Azure resources. We'll take the most userfriendly approach and do it through the **Azure portal**. When you are more experienced with Azure, you'll likely use any of the other approaches that are command- or code-based.

To create a Cognitive Services resource, you need to have access to an Azure subscription. If you don't have one yet, sign up for a free subscription at `https://azure.microsoft.com/en-us/free/`.

An *Azure subscription* is needed to create and pay for any resource within Azure. Multiple *resources* are grouped into *resource groups*, which provide an administrative grouping of resources created within a subscription.

After using an Azure subscription to create a Cognitive Services resource within a resource group, two things will be generated for you to consume the Cognitive Services resource, as outlined here:

- **Endpoint**: A **Uniform Resource Locator** (**URL**). To use a Cognitive Service, send data through a **HyperText Transfer Protocol** (**HTTP**) request to the URL, and the response will include the prediction made by the model you want to use (for example, get a description of an image).

- **Subscription key**: A code. To send data to the endpoint, you need to be authorized. To authorize, use one of the two generated subscription keys. You can regenerate any key at any time, which will disable the older version of the key.

Let's create a resource and obtain the endpoint and key through the following steps:

1. Open a browser and navigate to `http://portal.azure.com`.

2. Sign in with your account (the one with access to an Azure subscription).

3. In the menu on the left, select **+ Create a resource**, as shown in the following screenshot:

Figure 7.1 – Creating a resource

4. In the search bar that appears, as shown in the following screenshot, search for `cognitive services`:

Home >

Create a resource ···

Get started

Recently created

Categories

AI + Machine Learning

Analytics

Blockchain

Compute

Containers

🔍 cognitive services   ✕

Popular products See more in Marketplace

Windows Server 2019 Datacenter
Create | Learn more

Ubuntu Server 20.04 LTS
Create | Learn more

Web App
Create | Docs | MS Learn

Figure 7.2 – Searching for Cognitive Services

5. Select the **Cognitive Services** card that appears in the results, as shown in the following screenshot:

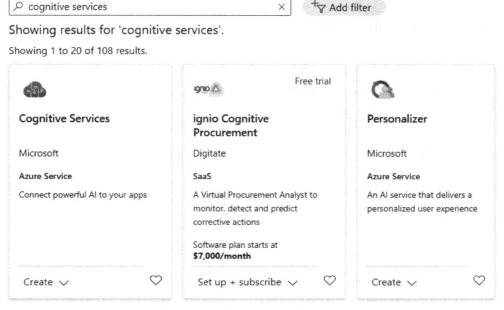

🔍 cognitive services   ✕   ⁺▽ Add filter

Showing results for 'cognitive services'.

Showing 1 to 20 of 108 results.

Cognitive Services	ignio Cognitive Procurement	Personalizer
	Free trial	
Microsoft	Digitate	Microsoft
Azure Service	SaaS	**Azure Service**
Connect powerful AI to your apps	A Virtual Procurement Analyst to monitor, detect and predict corrective actions	An AI service that delivers a personalized user experience
	Software plan starts at **$7,000/month**	
Create ⌄ ♡	Set up + subscribe ⌄ ♡	Create ⌄ ♡

Figure 7.3 – Cognitive Services card

6. After selecting the card, an overview of the Cognitive Services resource appears, as shown in the following screenshot. Select **Create**:

Home > Create a resource > Marketplace >

Cognitive Services ⚲ ⋯
Microsoft

Cognitive Services ♡ Add to Favorites

Microsoft

★ 4.3 (30 Azure ratings)

[Create]

Overview   Plans   Usage Information + Support   Reviews

Cognitive Services is a product bundle that enables customers to access multiple services with a single API key.

Product features:

Access to Vision, Language, Search, and Speech services using a single API

Quickly connect services together to achieve more insights into your content

Easily integrate with other services like Azure Search

Figure 7.4 – Creating a Cognitive Services resource

7. Select the **Subscription** option you want to use (with which you'll pay for the resource).

8. Choose an existing resource group or create a new one by selecting **Create new**.

9. Choose an available region closest to you.

10. Enter a unique **Name** value. You'll be notified immediately if the name is not unique.

11. Choose a **Pricing tier** option. Go for **Free F0** if available (when it's your first resource); otherwise, choose **Standard S0**.

    The following screenshot shows an overview of the configuration before creating a resource:

Project details

Select the subscription to manage deployed resources and costs. Use resource groups like folders to organize and manage all your resources.

Subscription * ⓘ

> Visual Studio Enterprise   ⌄

   └─── Resource group * ⓘ

> (New) ai-pbi-rg   ⌄
> Create new

Instance details

Region * ⓘ

> West Europe   ⌄

ⓘ Location specifies the region only for included regional services. This does not specify a region for included non-regional services. Click here for more details.   ⌐⅃

Name * ⓘ

> cs-pbi   ✓

Pricing tier * ⓘ

> Standard S0   ⌄

Figure 7.5 – Project details

12. Select **Review + Create**.

13. A summary of the project details is shown. Select **Create**.

    It should only take a couple of seconds for a resource to be created. Once the deployment is complete, your screen should look like this:

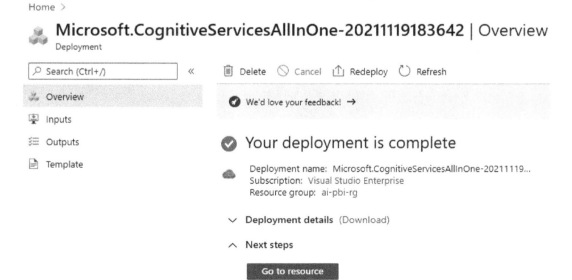

Home >

Microsoft.CognitiveServicesAllInOne-20211119183642 | Overview
Deployment

🔎 Search (Ctrl+/)   «   🗑 Delete   ⊘ Cancel   ⬆ Redeploy   ⟳ Refresh

🎚 Overview   ⊘ We'd love your feedback! →

🖥 Inputs

≋ Outputs   ✅ Your deployment is complete

🗎 Template   Deployment name: Microsoft.CognitiveServicesAllInOne-20211119...
                                Subscription: Visual Studio Enterprise
                                Resource group: ai-pbi-rg

                            ⌄ **Deployment details** (Download)

                            ⌃ **Next steps**

                                Go to resource

Figure 7.6 – Deployment complete

14. Select **Go to resource**.

The resource is now created and ready to be used. You should now be in the
Overview tab of the resource, as shown in the following screenshot. To use any of
the APIs, you'll need a key and the endpoint:

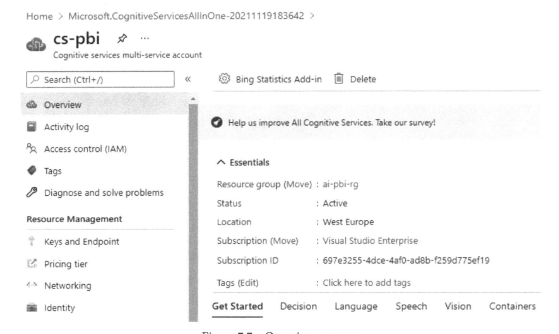

Figure 7.7 – Overview resource

15. In the menu on the left, under **Resource Management**, select **Keys and Endpoint**.

In the **Keys and Endpoint** screen, as shown in the following screenshot, you'll see
KEY 1, **KEY 2**, **Location/Region**, and **Endpoint**:

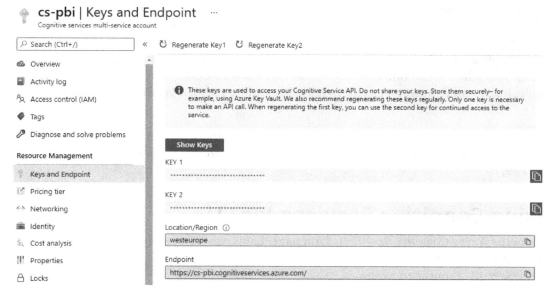

Figure 7.8 – Keys and Endpoint screen

KEY 1 and **KEY 2** have the same functionality. Any time you need to provide a **subscription key**, you can use either of these. You have two so that you can refresh one for security purposes, while still using the other, without breaking your application. Depending on which API you use, you either need to provide the region of your Cognitive Service or the endpoint, as we will experience when calling the API from Power BI.

We can integrate Cognitive Services with many kinds of applications, as you can call an API using various frameworks. For example, if you create an application with C# and want to automatically generate descriptions of your images, you can use the Computer Vision model provided by the Cognitive Services. To integrate the **Computer Vision** API with your application, you can use the **software development kit (SDK)** for C#.

Each Cognitive Service supports common languages such as C#, Node.js, and Python through these SDKs, making it easier to integrate an AI model. In the next chapters, we'll explore how we can integrate Cognitive Services with Power BI.

In this chapter, we'll use the Python framework as an example of calling the API through code to focus solely on the services for now.

All examples in this chapter run Python scripts from **Visual Studio Code** (**VS Code**). Why Python? Because, as discussed in previous chapters, this is often the preferred language of a data scientist. The full details of setting up VS Code and how to run these scripts are outside of the scope of this book. Nevertheless, if you want to set this up yourself, you can find instructions on GitHub at `https://github.com/ PacktPublishing/Artificial-Intelligence-with-Power-BI/tree/ main/Chapter07`.

We have created a Cognitive Services resource in Azure, and with this, we can make use of the prebuilt models Azure offers. In subsequent sections, we'll use this resource to analyze our data.

Using Cognitive Services for LU

One of the Azure *Cognitive Services* is **Language**. Language refers to any insights and meaning we can derive from text. Two features within this category are interesting to use with Power BI, as outlined here:

- **Text Analytics**: This includes many features that help extract information from unstructured text.

- **Question answering**: This creates a search engine on a knowledge base that contains mostly text that can be queried with natural language.

In the following sections, we will delve into the options we have to analyze text, and how we can use *question answering* to create an intuitive search engine from a knowledge base that resides outside of Power BI.

Using Azure's Text Analytics

Text can be easy for us to read but is hard for data analysts to work with. As part of your data, you may have product reviews or feedback forms that are challenging to visualize in Power BI. To extract some key insights from unstructured text such as reviews or forms, we can use several **natural LU** (**NLU**) methods.

Imagine you're planning a holiday, and you're using a website to browse through available hotels at your destination. For each hotel, you not only look at the description but also at the reviews of other guests. You are specifically interested to learn whether a hotel has a good breakfast. Instead of reading through all the reviews, you want to be able to filter on a key *phrase* such as *breakfast* and see whether the review is *positive* or *negative*.

Analyzing text to extract **key phrases** and detecting **sentiment** are two features of the Cognitive Services for language. Some other features are outlined here:

- **Personally Identifiable Information (PII) detection**: Detect PII data that you want to flag as sensitive information. You may want to detect PII data to ensure it's masked or deleted from the data.

- **Named Entity Recognition (NER)**: A form of key-phrase extraction where the key phrases are categorized as—for example—a person, organization, or location.

- **Entity linking**: An extracted key phrase or entity may refer to different meanings. For example, *Holland* can refer to Tom Holland the actor or Holland the country. To disambiguate, a link to the relevant Wikipedia page can be provided with an entity.

To learn more about the other features, read the documentation at `https://docs.` `microsoft.com/en-us/azure/cognitive-services/language-service/` `overview`.

By extracting key phrases such as *breakfast* and sentiments such as *positive* or *negative* from hotel reviews, you can create richer visualizations in Power BI. For example, you can visualize which features of your hotel are often mentioned in positive reviews, giving you insights into what customers like (or dislike) about your hotel; or, you can see how the sentiment score in hotel reviews changes over time. By analyzing the sentiment over time compared to changes made, such as changing the breakfast options, you can see what the effect of changes is on customer satisfaction. In other words, extracting insights from unstructured data such as text will help you uncover more information.

To extract insights from text, we'll use one of the Language APIs. We'll need a Cognitive Services resource to consume any Language API. In the previous section, we learned how to create a Cognitive Services resource using the Azure portal. Since all we need to do is call the API, we can run a simple Python script from our local system to do this.

Using VS Code, we can run a Python script that includes the following steps:

1. Connects to the API using the subscription key and endpoint of a previously created Cognitive Services resource

2. Loads a locally stored folder with text files

3. Sends the text files to the Text Analytics API and returns a list of key phrases and the overall sentiment score

Building on the example of hotel reviews, we'll explore how we can use the Language API to extract insights from five short customer reviews on hotels. The hotel reviews and a Python script you can use to extract insights can be found on GitHub at https://github.com/PacktPublishing/Artificial-Intelligence-with-Power-BI/tree/main/Chapter07/language.

To get an idea of how the Language API works, let's see what the code could include if we were to use the Python SDK for the Language API. First, we want to connect to the Cognitive Services resource by using one of the authentication or subscription keys and the endpoint or URL created for us, as follows:

```
# Connect to API through subscription key and endpoint
subscription_key = "<your-subscription-key>"
endpoint = "https://<your-cognitive-service>.cognitiveservices.
azure.com/"

# Authenticate
credential = AzureKeyCredential(subscription_key)
cog_client = TextAnalyticsClient(endpoint=endpoint,
credential=credential)
```

After authentication, you need to load the data you want to send to the Language API. In this case, we have text files stored in a local folder. All we need to do is refer to the folder, as follows:

```
# Get reviews
reviews_folder = 'reviews'
```

Finally, you specify what you want the Cognitive Service to do. In the following code example, we want to extract the key phrases and the sentiment from the text file that we send as input:

```
    # Get key phrases
    phrases = cog_client.extract_key_phrases(documents=[text])
[0].key_phrases
    if len(phrases) > 0:
        print("\nKey Phrases:")
        for phrase in phrases:
            print('\t{}'.format(phrase))

    # Get sentiment
```

```
    sentimentAnalysis = cog_client.analyze_
sentiment(documents=[text])[0]
    print("\nSentiment: {}".format(sentimentAnalysis.
sentiment))
```

The script refers to the `reviews` folder, which contains three text files—each a short review of a hotel. After running the code, we get the following results:

- The input and output for `review1.txt`, as illustrated here:

```
I had an excellent stay, thanks so much! The only
feedback I have is that the pillow was too hard.
Key Phrases:
        excellent stay
        feedback
        pillow


Sentiment: mixed
```

- The input and output for `review2.txt`, as illustrated here:

```
I didn't really like the breakfast. There were very few
options to choose from and it didn't feel like a healthy
start to the day.
Key Phrases:
        healthy start
        breakfast
        options
        day


Sentiment: negative
```

- The input and output for `review3.txt`, as illustrated here:

```
We had the best time here. The owner is very friendly and
went out of his way to help us. Couldn't have wished for
a better welcome to Croatia.
Key Phrases:
        best time
        owner
        way
```

```
        welcome

        Croatia


Sentiment: positive
```

The preceding examples show how we can submit pieces of text to the Text Analytics API to extract key phrases and sentiment. As mentioned earlier, there are other insights you can extract from such texts—for example, you can detect the language of the text. To learn more about how to implement the different features, go to the API reference documentation at https://docs.microsoft.com/en-us/azure/cognitive-services/language-service/overview.

Notice how quickly we can extract these insights from text. We don't need to have extensive data science expertise to train a sophisticated NLU model to analyze text. Instead, we can simply send our data to the API and get the required insights back.

Unfortunately, there is a downside to using prebuilt models. For the Text Analytics API, we don't know exactly which NLU models support it. From common NLU approaches, we can assume that **deep learning** (**DL**) algorithms are used, not just because text is unstructured data, but also because to understand text, we have to understand the words as well as their context. However, for each Text Analytics feature, it is likely that different models are used.

We have experienced how we can use an Azure Cognitive Service for language to extract key phrases and sentiment from text. How we can use these insights in Power BI reports will be shown in the next chapter, *Chapter 8*, *Integrating Natural Language Understanding with Power BI*.

Creating question answering from a knowledge base

Another feature in the Language category that can be interesting for us is **question answering**. Question answering enables us to create a **question-and-answering** (**Q&A**) layer on top of a knowledge base. With this service, users can ask a question using natural language to get an answer from our knowledge base.

As you may recall from *Chapter 6*, *Using Natural Language to Explore Data with the Q&A Visual*, Power BI already has a Q&A visual we can use to help the end user of our reports, by letting them ask questions about our data and get immediate answers that are extracted from the data.

In addition to the Q&A visual, we can also integrate question answering with Power BI to make it possible for end users to ask any **frequently asked questions (FAQs)** they may have; for example, they may want to know how often the data is refreshed or where it comes from. You can store all this information in a knowledge base somewhere outside of Power BI and use question answering to create a Q&A service that you can then integrate with Power BI.

In *Chapter 9*, *Integrating an Interactive Question and Answering App into Power BI*, we will learn how to create a service and integrate it with Power BI. After that chapter, you'll be able to add this feature to your Power BI reports so that you can make any extra information easily accessible to your end users.

Using Cognitive Services for CV

As well as adding images to your Power BI reports and dashboards, you can add images as data. To extract insights from images that you can visualize in your report, you can use prebuilt Computer Vision models. Depending on the type of images you have and the information you want to extract, you can use different APIs of the Azure Cognitive Services.

The three main Azure Cognitive Services you'll use with Power BI are listed here:

- **Computer Vision**: Use prebuilt models to describe or categorize images based on generic models
- **Custom Vision**: Customize a partially prebuilt model by adding your own data to categorize images or detect objects in images
- **Face**: Detect and analyze human faces and facial features in images

Let's explore each of these three CV services.

Understanding Azure's Computer Vision

Although **Azure's Computer Vision** is the service that analyzes images, *CV* is also a term in and of itself. We'll go over both the term as well as the service, to understand how and when to use the service.

CV refers to the field of AI that includes models trained in extracting information from images. Compared to tabular data, images are much harder to analyze and contain a lot of information.

Next to that, the challenge with images is that we want to train AI to look at images the way that we do. For example, when you're about to cross a road, there are many things that you can see, but the most important thing you want to perceive is whether a car is near you. It doesn't matter that there is a garbage can next to you—the focus should be on the car. With CV, we're trying to accomplish a similar so-called focus when looking at images. We want to extract the most important information that we humans would consider important.

What we consider important, of course, depends on why we're interested in an image. That is also why CV has distinctive subfields—for example, it depends on whether you want to see if you're looking at an image of a dress *or* a T-shirt, or if you want to see *where* in the image a car is. The former is what we can achieve with **image classification**, while the latter is an example of **object detection**. We'll discuss these two subfields in more detail in the next section when exploring Custom Vision.

The reason it is important to understand the many different subfields of the CV field is that it also relates to the many different features of *Azure's Computer Vision*. Which feature you use naturally depends on the use case at hand. For a full list of what the Computer Vision API can do, explore the API documentation at `https://westus.dev.cognitive.microsoft.com/docs/services/computer-vision-v3-2/operations/56f91f2e778daf14a499f20d`.

Knowing that our goal is to visualize extracted information from images in Power BI, some interesting Computer Vision API features are outlined here:

- Detecting mature content
- Categorizing the content of an image according to a predefined taxonomy
- Describing an image with a sentence
- Detecting objects in an image
- Returning a list of words or tags related to the image content
- Identifying celebrities and landmarks
- Detecting text in an image, also known as **optical character recognition** (**OCR**)

Although no information is provided about the underlying models, it is safe to assume that DL techniques are used. The use of **neural networks** (**NNs**) in DL that resemble the mechanisms of the human brain has tremendously helped the field of CV to *focus* on what is needed in images. Training DL models that can analyze images is a complex and time-consuming undertaking, and well beyond the scope of this book. The complexity, however, is also a strong argument to use prebuilt models instead of investing in creating your own, especially if you don't already have the resources.

Now that we know what the field of CV is and what Azure Computer Vision can do for us, let's see it in action in the next section.

Using Azure's Computer Vision

As discussed previously, Azure Computer Vision has many features that can be of benefit to us. We can use the API to analyze images and extract information to visualize in our Power BI reports. To get an idea of what Azure Computer Vision can do, let's look at some examples of consuming the API.

After creating a Cognitive Services resource through the Azure portal, as shown in the first section of this chapter, you can obtain an endpoint and key to use any of the APIs. For a complete tutorial on using the Computer Vision API with a variety of programming languages, read the *Quickstart* page at `https://docs.microsoft.com/en-us/ azure/cognitive-services/computer-vision/quickstarts-sdk/image- analysis-client-library?tabs=visual-studio&pivots=programming- language-python`.

To use the Computer Vision API when working with Python, create a script that does the following:

1. Connects to the API using the subscription key and endpoint of a previously created Cognitive Services resource

2. Loads a locally stored image

3. Sends the image to the Computer Vision API and returns a description

To explore the Computer Vision API in this chapter, we'll see how we can use the API to extract insights from images that show objects found inside the house or outside on the street.

First, we'll use the Computer Vision API to describe the image for us. Find an example of a Python script that will return a description of images, `describe-image.py`, and sample data on GitHub at `https://github.com/PacktPublishing/ Artificial-Intelligence-with-Power-BI/tree/main/Chapter07/ vision`.

To connect to the API, you can define variables for the key and endpoint with the values you retrieve from the Cognitive Services resource, as follows:

```
# Connect to API through subscription key and endpoint
subscription_key = "<your-subscription-key>"
endpoint = https://<your-cognitive-service>.
cognitiveservices.azure.com/


# Authenticate
computervision_client = ComputerVisionClient(endpoint,
CognitiveServicesCredentials(subscription_key))
```

Next, you load an image stored locally, which you then send to the API by using the Python SDK. Here's the code you'll need in order to do this:

```
# Get image
local_image = open("images/street.jpeg", "rb")


# Call API
description_result = computervision_client.describe_image_in_
stream(local_image)
```

And finally, you can return the result of the API call and print it so that you can view the results, as follows:

```
# Return the result
print("Description of local image: ")
if (len(description_result.captions) == 0):
    print("No description detected.")
else:
    for caption in description_result.captions:
        print("'{}' with confidence {:.2f}%".format(caption.
text, caption.confidence * 100))
print()
```

The code used in the Python script refers to a street.jpeg image that is sent to the Computer Vision API; this is shown in the following screenshot:

Figure 7.9 – street.jpeg

After running the script on the `street.jpeg` file, we get the following result:

```
Description of local image:
'cars parked in front of a building' with confidence 53.42%
```

Earlier, we discussed how we want a Computer Vision model to focus on the most important objects in an image. Using again the example of crossing the road, the most important objects for us are the cars we see in the image. Even though there are trees in the background, these can be considered less important than cars, which is why the trees are not mentioned in the description.

Still, we may want to detect all objects that are present in an image. For that, we can also use the object detection feature of the Computer Vision API. Again, we'll have to use the endpoint and subscription key of the Cognitive Services resource, after which we can send the image to the Computer Vision API and define how we want the result to be returned. An example of a complete `detect-objects.py` Python script can be found on GitHub at `https://github.com/PacktPublishing/Artificial-Intelligence-with-Power-BI/tree/main/Chapter07/vision`.

After authentication, the call to the Computer Vision API with Python could look like this:

```
# Call API
detect_objects_results_local = computervision_client.detect_
objects_in_stream(local_image)
```

After calling the API, you can return the result with the following `print` execution:

```
# Return the result
print("Detected objects in image:")
if len(detect_objects_results_local.objects) == 0:
    print("No objects detected.")
else:
    for object in detect_objects_results_local.objects:
        print(object.object_property)
print()
```

The local image that is sent to the Computer Vision API, in this case, is the one shown here:

Figure 7.10 – Objects

After running the code, the following objects are detected in the image:

```
Detected objects in image:
'Luggage and bags' with confidence 73.80%
'potted plant' with confidence 63.30%
'dog' with confidence 93.40%
```

This example shows the power of the Computer Vision API. As we have experienced, it includes various features that can help us analyze images. Which feature you use depends on your use case.

Now that we have known what Computer Vision can do, let's take it one step further with Azure's Custom Vision.

Using Azure's Custom Vision

A service such as Computer Vision is great as the prebuilt model can immediately be used and requires no extensive data science expertise. Still, there are situations where you do need a model that is catered to your use case. For those scenarios, we can use Azure's Custom Vision.

For example, in *Figure 7.10*, the objects were detected as luggage and bags, potted plant, and dog. If you are creating an application that is supposed to detect plants in images to give the user information on how to take care of it, you also want to know what kind of plant this is. Computer Vision has its own taxonomy and will not offer that level of detail. To achieve that, you can *further* train a model by using Custom Vision.

Training your own model may sound daunting, but it isn't with Custom Vision. The service has been created to be easy to use, even for people with little to no data science knowledge. Instead of creating a Computer Vision model from scratch, you're already using a partially trained model that is sitting underneath the surface. By adding only a small number of your own images, you can train a Computer Vision model more catered to your needs.

There are two types of models you can create with Custom Vision, as outlined here:

- **Image classification**: By providing labeled images, a complete image is classified as the label you provide.

- **Object detection**: By providing images with labeled bounding boxes, objects within images are detected.

> **A model only knows what it is taught**
>
> A custom vision model will only know what it is taught. The labels you provide will be the only labels it can predict. Even though there is an already partially trained model underneath, it can't extrapolate insights based on labels entered by you.

To use Custom Vision, you have to add images through the Custom Vision portal or by using the SDK. We will cover this in *Chapter 10, Getting Insights from Images with Computer Vision*.

For now, we have learned that Custom Vision will enable us to customize a Computer Vision model for image classification or object detection use cases.

Using the Face service

The Computer Vision API can already detect faces. In some cases, however, we may want more information than that. Using the **Face service**, you can extract facial features to get more detailed information about the faces you are analyzing.

The Face API provides many features that involve the face, such as face identification and verification. Some of these features are likely not to be of use when working with Power BI. That is why we will focus here on the **face detection and attributes** feature.

Next to simply detecting faces with the Face API, we can also detect things such as these:

- Age
- Emotion
- Glasses
- Hair
- Mask

For example, as an organization, you may have a security protocol in place where it is required to wear glasses for safety. By monitoring the workplace, you can detect faces and detect whether they are wearing glasses. Subsequently, you can visualize in a Power BI dashboard how many people are wearing glasses, and thus how well people keep to the security protocols.

The Face service is one of those prebuilt models that can be very complex to build yourselves, especially when trying to ensure there is no bias in the data or the model. Another risk with the Face service is in the application of it. Although there may be interesting applications, they shouldn't always be pursued because of privacy and human rights concerns. Some guidance into how to use AI responsibly is provided in *Chapter 13, Responsible AI*.

You can learn more about how to use the Face API for face and attribute detection from the following documentation: `https://docs.microsoft.com/en-us/azure/cognitive-services/face/concepts/face-detection`.

Although you should be cautious about processing images containing faces, you can detect facial features with the Face API. The insights from this service can then be visualized in a Power BI report.

Summary

We have had our first introduction to Azure's Cognitive Services in this chapter, a suite of APIs that enable us to use prebuilt models and integrate them with any application, such as Power BI. We have learned about how to use Cognitive Services for language to extract key phrases and detect sentiment, as well as create a custom question-answering service. We have also learned how to use the Cognitive Services for vision, which enables us to extract information from images.

In the next chapter, we'll see how we integrate Azure Cognitive Services for language with Power BI to visualize insights from data containing text.

8

Integrating Natural Language Understanding with Power BI

Whether you are working with small or large amounts of text, it is challenging to automate the ability to extract insights. Imagine customers leaving you with product reviews. By recognizing trends or common issues, you know how to improve the product.

To analyze unstructured text, you can use prebuilt models from Azure Cognitive Services. You can easily integrate these models with Power BI. After using **natural language understanding** (NLU) models to extract data from text, you can visualize the insights using a Word Cloud.

In this chapter, we'll cover two main ways to call the Language APIs in Power BI Desktop, after which we'll visualize the insights to create a Power BI report. To do this, we will be covering the following topics:

- Using Language APIs in Power BI Desktop
- Visualizing insights from text in reports

Upon completing this chapter, you'll be able to analyze unstructured text in Power BI by using prebuilt models.

Technical requirements

There are a couple of things you need to go through the examples provided in this chapter:

- Power BI Desktop:

  As this book revolves around Power BI, we expect you to have Power BI Desktop installed on your system. You can install Power BI from the Microsoft Store or find more advanced download options here: `https://www.microsoft.com/en-us/download/details.aspx?id=58494`.

- A Cognitive Services resource:

  Cognitive Services provides resources through Azure. To create a resource, you need to have access to an Azure subscription. If you don't have one yet, you can sign up for a free subscription here: `https://azure.microsoft.com/en-us/free/`.

  If you don't have a Cognitive Services resource, you can create one using your Azure subscription. For instructions on how to do so, go to *Chapter 7, Using Cognitive Services*.

- Twitter sample dataset:

  The first sample dataset that we'll be using includes tweets that mention *artificial intelligence*. The data is stored as a JSON file and can be downloaded from GitHub: `https://github.com/PacktPublishing/Artificial-Intelligence-with-Power-BI/tree/main/Chapter08/aitweets.json`.

- Hotel reviews sample dataset:

  The second sample dataset contains reviews that have been posted for different hotels in Europe. The CSV file can be downloaded from GitHub: `https://github.com/PacktPublishing/Artificial-Intelligence-with-Power-BI/tree/main/Chapter08/hotel_reviews.csv`.

Let's start exploring the NLU features in Power BI!

Using Language APIs in Power BI Desktop

Whether you are working with small or large amounts of text, it can be challenging to extract interesting insights from text that can be visualized in Power BI.

Depending on the text you have and the reason you want to analyze it, there may be different insights you want to extract from it. The three most common insights that are extracted from text are as follows:

- **Key phrases**: To find the main concepts in a piece of text
- **Sentiment labels**: To determine whether the text is positive or negative
- **Language**: To detect which language the text is written in

In this section, we'll focus on these insights to explore how they can be extracted from text in Power BI, and how those insights can be visualized. The same approach can be used for any other insights you may want to derive from text by using other models or APIs.

Once you have loaded data into Power BI that contains text, the easiest way to derive insights from the column containing the text is to use the **Cognitive Services Language API**. There are two main approaches to integrating the Language API with Power BI Desktop:

- **Using AI Insights**: This is the easiest option to use but requires a Premium license.
- **Using Power Query Editor**: This involves more configuration and requires an Azure subscription.

Send No Empty Rows to the Language APIs

Whenever you use the Language APIs, the text field that will be analyzed should not have any empty rows. If your dataset contains empty rows, clean your data by, for example, using the **Remove Empty** filter to make sure the contents of these rows will not be sent to the Language API.

Now that we know what insights we want to extract from the text field in Power BI, let's explore the two approaches we can take.

Using AI Insights

One way you can call upon an Azure Language API from Power BI is through the AI insights feature. The AI Insights feature is built into Power BI and is the easiest way to make use of the Azure Cognitive Services Language API. To use it, we simply have to use a button that will guide us through the process.

Let's try it out with some text data. For this, we'll use the Twitter dataset, which contains tweets over a period that mention `artificial intelligence`. The data is semistructured as it contains fields containing metadata for each tweet, as well as the tweet's text.

The Twitter dataset that we'll be using can be downloaded from `https://github.com/PacktPublishing/Artificial-Intelligence-with-Power-BI/tree/main/Chapter08/aitweets.json`. The dataset should be stored locally as `aitweets.json`. Let's get started:

1. Open a new Power BI report in Power BI Desktop.

2. From the top ribbon, under the **Home** tab, select **Get data**, as shown in the following screenshot:

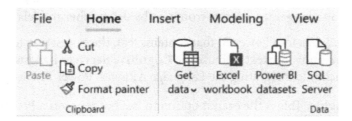

Figure 8.1 – Get data

3. On the pop-up screen for **Get Data**, select **JSON**, as shown in the following screenshot:

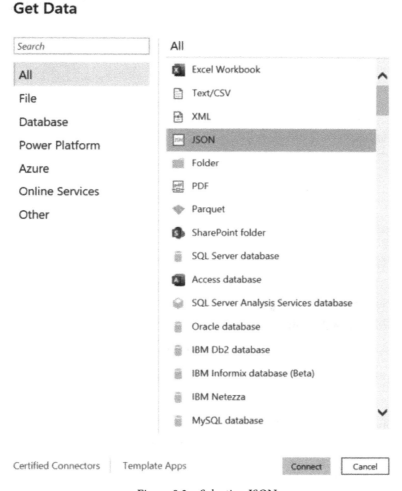

Figure 8.2 – Selecting JSON

4. Select **Connect**.

5. Navigate to the `aitweets.json` file and click **Open**.

6. In the **Connection settings** window, select **Import**, and then select **OK**.

The data should now be imported into Power BI, as shown in the following screenshot:

	CreatedAt	1²₃ RetweetCount	ᴬᴮc TweetLanguage	ᴬᴮc TweetText
1	7/14/2021 14:47:39	0	en	@____bruvteresa_ According to resea...
2	7/14/2021 14:47:49	2	en	RT @HDataSystems: Artificial Intellige... #hdatasystems #Artificia...
3	7/14/2021 14:47:33	246	en	RT @adgpi: Army Technology Board c...
4	7/14/2021 14:47:35	1	en	RT @pacorjo: According to a recent su...
5	7/14/2021 14:47:54	20	en	RT @HarbRimah: Making AI Sing https... #MachineLearning #DataScience #Python
6	7/14/2021 14:48:35	1	en	RT @weblineglobal: The applications o...
7	7/14/2021 14:48:36	1	en	RT @sokoworlddotcom: We help our ...
8	7/14/2021 14:48:37	1	en	RT @SuriyaSubraman: US Falling Behi...
9	7/14/2021 15:48:37	2	en	RT @DD_FaFa_: Prediction Machines: ...
10	7/14/2021 15:48:27	1	en	RT @williamhersh: Latest post to the l...

Figure 8.3 – Twitter data

The `aitweets` table includes the following fields:

- `CreatedAt`: The date and time when the tweet was posted
- `RetweetCount`: How many times the tweet was retweeted
- `TweetLanguage`: The language of the tweet, as detected by Twitter
- `TweetText`: The text of the tweet

You're probably already familiar with working with date and number variables in Power BI reports. A tweet text, however, is a bit challenging to visualize. A quick way to extract insights that we can use for visualizations is to use the AI Insights that are already in Power BI. Let's learn how to use these AI Insights.

7. From the **Home** ribbon, select the **Text Analytics** feature from the **AI Insights** category. You can find it at the top right:

Figure 8.4 – Text Analytics

After selecting the **Text Analytics** feature, a pop-up window will appear, as follows:

Text Analytics

▲ Text Analytics [3]

  fx Detect language

  fx Extract key phrases

  fx Score sentiment

Extract key phrases

Sift through text and surface important phrases in your data.

Learn more

Text

| ▼ | TweetText | ▼ |

Language ISO code (optional)

| A^B_C ▼ | Example: abc |

Premium capacity used for AI Insights

Default (based on availability)   ▼

OK   Cancel

Figure 8.5 – The Text Analytics window

There are a couple of things to note in this pop-up:

- In the left pane, there are the three **Text Analytics** features we can use in Power BI: detecting the language of the text, extracting key phrases, and determining the sentiment of a piece of text.

- At the bottom left, we can see which **Premium capacity** is being used for AI Insights. You can choose which capacity you want to use for this feature by using the drop-down menu here.

- To the right, we can see the feature we selected from the **Text Analytics** options and configure it. The function needs to know which field to analyze and send to the Text Analytics API. Optionally, we can tell it what language the text is written in. By default, it will automatically try to detect the language, but if you want to overrule this, you can do so by providing the ISO code of the language you're using (for example, en is the ISO code for English).

8. In the **Text Analytics** pop-up window, select **Extract key phrases** from the left menu.

9. In the right pane of the pop-up window, select **TweetText** as the input field for **Text**.

10. Select **OK**.

You will see five yellow dots moving horizontally across the Power BI Query Editor, which is a sign of the data model loading. After you invoke the Text Analytics function, Power BI will send your data to the Text Analytics API and get the extracted key phrases in return, as shown in the following screenshot:

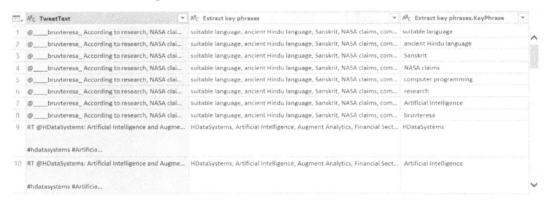

Figure 8.6 – Extracted key phrases

In the preceding screenshot, two new columns have been added by the Text Analytics feature:

- **Extract key phrases**: A list of the key phrases that have been extracted from the text, separated by commas.

- **Extract key phrases.KeyPhrase**: From the list, each key phrase is extracted and given a row, making it easier to use the data for Power BI visualizations. However, this means that the rows will have been duplicated as the same tweet will have been copied for each key phrase that has been extracted from the tweet.

Once Power BI is connected to data, key phrases will be extracted with just a couple of clicks. Later in this chapter, we will learn how to visualize the newly acquired data.

Note that to detect the language, or to determine the sentiment, we go through the same steps. All we have to do is choose a different Text Analytics feature from the **Text Analytics** pop-up window. When you're detecting a language, two columns will be added:

- **Detected Language Name**: The full name of the language that's been detected; for example, Spanish

- **Detected Language ISO Code**: The ISO code of the language that's been detected; for example, es

When you want to see what the sentiment of a piece of text is, you will get one extra column:

- **Sentiment score**: The sentiment of the text, which will be valued as either positive, negative, mixed, or neutral

For more information, read the documentation on how to use AI Insights in Power BI Desktop: https://docs.microsoft.com/en-us/power-bi/transform-model/desktop-ai-insights.

Using AI Insights for **Text Analytics** is the quickest and easiest option to detect language, extract key phrases, and determine their sentiment. However, you do need Power BI Premium to use the feature. So, let's explore an alternative.

Using Power Query Editor

For more control and flexibility, we can call an API from Power Query Editor. This approach can be used for many different APIs and allows you to use the various features an API offers (for example, extracting key phrases or detecting sentiment).

For this example, we will use another dataset. This time, we will extract useful information from hotel reviews. You can download the dataset from https://github.com/PacktPublishing/Artificial-Intelligence-with-Power-BI/tree/main/Chapter08/hotel_reviews.csv. In the following steps, the data will be referred to as hotel_reviews.csv:

1. Open a *new* Power BI report in Power BI Desktop. You mustn't use the AI functions in this report yet. Otherwise, you will get an error when invoking the custom function later.

2. From the top ribbon, under the **Home** tab, select **Get data**.

3. Select **Text/CSV**.

4. Navigate to where you stored hotel_reviews.csv, select the file, and **Open** it.

5. After reviewing the preview of the data, select **Transform data** to open **Power Query Editor**.

6. Under the **Home** ribbon of **Power Query Editor**, select **New Source**.

7. Select **Blank Query**, as shown in the following screenshot:

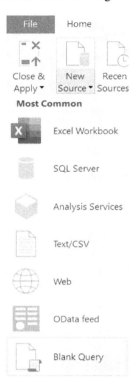

Figure 8.7 – Blank Query

In the **Queries** pane on the left, you should now have a second query called Query1. We will use this query to create a function that will make the API call to the **Text Analytics** service.

8. Select Query1 from the **Queries** pane on the left.

9. Select **Advanced Editor** from the **Home** ribbon. If the option is grayed out, you may have to click on the formula bar first.

A pop-up window will appear, as shown in the following screenshot. This window is the **Advanced Editor** window and is where you can edit Query1:

Figure 8.8 – Advanced Editor

10. Delete the current content.

11. Copy and paste the following code. You can also find this code at https://github.com/PacktPublishing/Artificial-Intelligence-with-Power-BI/tree/main/Chapter08/customfunctions.md:

```
// Returns key phrases from the text in a comma-separated
list
(text) => let
    apikey      = "YOUR_API_KEY_HERE",
    endpoint    = "https://<your-custom-subdomain>.
cognitiveservices.azure.com/text/analytics" & "/v3.0/
keyPhrases",
    jsontext    = Text.FromBinary(Json.FromValue(Text.
Start(Text.Trim(text), 5000))),
```

```
    jsonbody    = "{ documents: [ { language: ""en"", id:
""0"", text: " & jsontext & " } ] }",
    bytesbody   = Text.ToBinary(jsonbody),
    headers     = [#"Ocp-Apim-Subscription-Key" =
apikey],
    bytesresp   = Web.Contents(endpoint,
[Headers=headers, Content=bytesbody]),
    jsonresp    = Json.Document(bytesresp),
    keyphrases  = Text.Lower(Text.
Combine(jsonresp[documents]{0}[keyPhrases], ", "))
in  keyphrases
```

12. Replace YOUR_API_KEY_HERE with the subscription key of your Cognitive Services resource. Leave the quotation marks in!

13. Replace <your-custom-subdomain> with the name of your Cognitive Services resource. Leave the rest of the endpoint as-is. The window should look similar to the following:

Figure 8.9 – Edited query

14. Select **Done**.

15. As shown in the following screenshot, the function has been created and expects **text** as input:

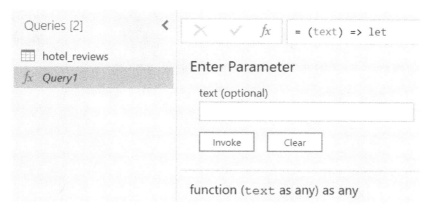

Figure 8.10 – New function

16. Rename Query1 for usability. Change it to Extract key phrases.

17. Select the hotel_reviews query.

18. From the **Add Column** tab of the top ribbon, select **Invoke Custom Function**, as shown in the following screenshot:

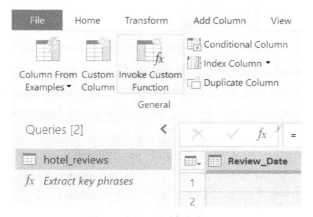

Figure 8.11 – Add Column

A pop-up window will appear where you can configure which custom function you want to invoke, as follows:

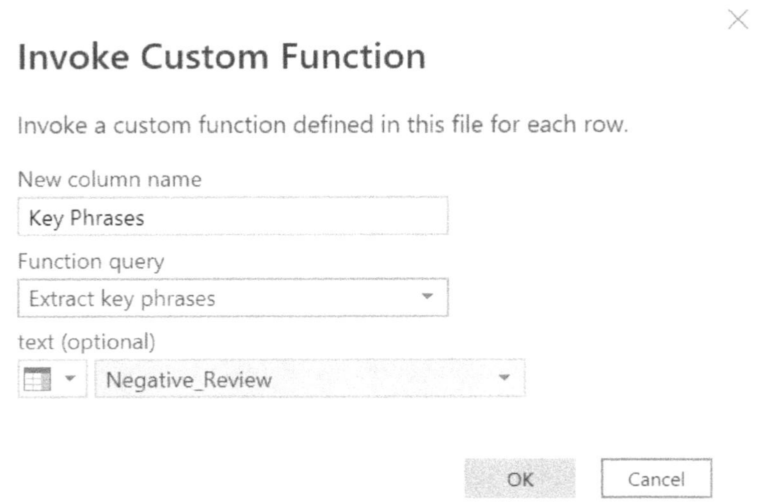

Invoke Custom Function

Invoke a custom function defined in this file for each row.

New column name

Key Phrases

Function query

Extract key phrases

text (optional)

Negative_Review

OK   Cancel

Figure 8.12 – Invoke Custom Function

19. For **New column name**, enter Key Phrases.

20. Select the Extract key phrases query you created for **Function query**.

21. For **text (optional)**, select Negative_Review.

22. Select **OK**.

Since we are making an API call to an external service, we have to let Power BI know how to connect to it. After selecting **OK**, in the **Invoke Custom Function** window, you will see a banner, asking you to specify how to connect. If the banner doesn't appear, you can skip the next two steps.

23. Select **Edit Credentials** to open a new pop-up window.

Power BI wants to know how it should authenticate for the **Extracting key phrases** query. Power BI doesn't know that we've already provided a key in the query, which is how we authenticate Power BI to use the Cognitive Services resource. To avoid Power BI adding an extra and unnecessary layer of authentication that will block the API call, we will set the authentication to **Anonymous**.

24. In the pop-up window that appeared, select **Anonymous** and select **Connect**.

You may get another banner that either tells you that information is required about data privacy or that the privacy levels do not work together. Either way, we need to make sure that the privacy levels of both the `hotel_reviews` query and the `Extract key phrases` function are set to **Public**.

> **Public Access to Data**
>
> The query that contains the API call must be set to public. Public data sources are visible to anyone with access to the Power BI report. If you are working with sensitive data, this approach causes a potential risk as unauthorized people can access the data. However, if you must set the data to private, you can work with **Azure Machine Learning (Azure ML)** to integrate similar artificial intelligence models. You will learn how to integrate with a model from Azure ML in *Chapter 12, Training a Model with Azure Machine Learning*.

Once the authentication and privacy blockers have been solved, a new column will be added to your `hotel_reviews` query, as shown in the following screenshot:

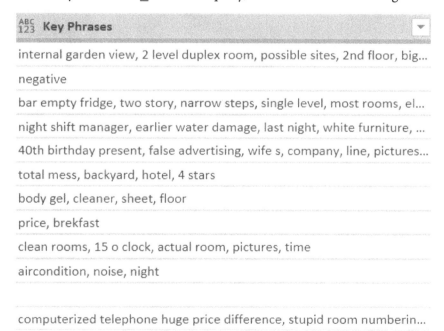

Figure 8.13 – Key Phrases from hotel_reviews

The new `Key Phrases` column contains the extracted key phrases for each row. The more rows you have, the longer it will take to process the data. The key phrases are separated with a comma, and Power BI will understand this when we want to visualize the individual key phrases, as we will explore in the next section.

Also, note that even though the data field that we extracted the key phrases from had no empty rows, there may be empty rows in this new column. Even when there is text, the Language API may not detect any key phrases in the text, in which case it will return nothing. An example of this is shown in the 11th row of the preceding screenshot.

Whatever approach you choose, the goal is to add a new column to your dataset that contains the insights you have extracted from the text field. Depending on your requirements and possible investment, you can use the AI Insights feature or Power Query Editor in Power BI to call a Language API. Subsequently, we would want to visualize those insights. Let's look at a way to put data consisting of text in your Power BI reports.

Visualizing insights from text in reports

Once you have enriched your dataset with insights that have been extracted from text, it is time to create the visuals. When you're working with insights from text, such as key phrases and sentiment, you can use many of the standard visuals. Now, let's learn what a Word Cloud is and how it can help visualize insights from text.

Visualizing text with a Word Cloud

A **Word Cloud** is a visual representation of words and their frequency. In other words, it visualizes words that occur often in our dataset. The more often they occur, the larger the word is represented in the visual.

By visualizing extracted key phrases with a Word Cloud, we can get intuitive insights into the most important terms in a dataset. Let's try it out with the key phrases we extracted from the hotel reviews dataset:

1. From the **Visualizations** pane, select the *three dots* icon to open the extra options menu, as shown in the following screenshot:

Figure 8.14 – Visualization options

2.   Select **Get more visuals**. A pop-up window called **Power BI visuals** will appear, as shown in the following screenshot:

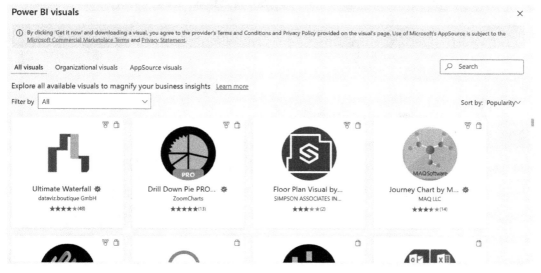

Figure 8.15 – Power BI visuals

3. At the top right, search for word cloud. As a result, you should see the **Word Cloud** card appear, as shown in the following screenshot:

Figure 8.16 – Search result

4. Select the **Word Cloud** search result.

5. After reviewing the summary information that's shown in the following screenshot, select **Get It Now**:

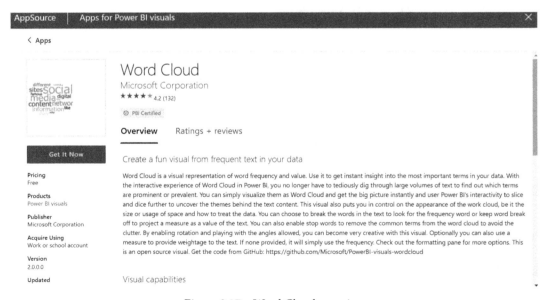

Figure 8.17 – Word Cloud overview

6. From the pop-up that appears, select **OK**:

Import successful

The visual was successfully imported into this report.

OK

Figure 8.18 – Import successful

7. The **Word Cloud** visualization should have been added to the bottom row of the **Visualizations** pane, as shown in the following screenshot:

Figure 8.19 – Word Cloud added to Visualizations

We are now ready to create a Word Cloud!

8. Add a **Word Cloud** visualization to your report by selecting the respective icon from the **Visualizations** pane.

9. From the hotel_reviews query, select Key Phrases and drag-and-drop it into the **Category** field in the **Visualizations** pane.

The following screenshot shows the Word Cloud that should have been added to your report:

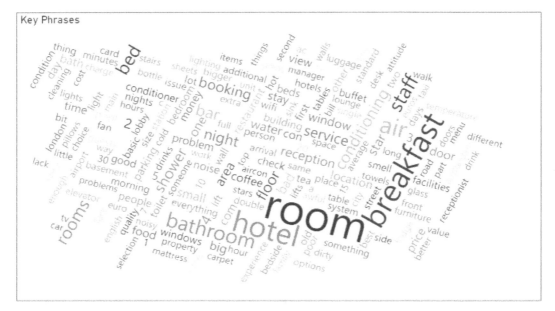

Figure 8.20 – Key phrases visualized in Word Cloud

The bigger the word in this Word Cloud, the more often it is mentioned in the negative hotel reviews. For example, here, we can see that many people have complained about the room and breakfast.

If you want to change what makes a word bigger in this visual, you can add the variable that should be used instead of the frequency (which is taken by default) to the **Values** field.

If you're not completely satisfied with the way the Word Cloud looks, you can edit its format. Let's take a look.

10. When **Word Cloud** is selected, go to the **Format** tab in the **Visualizations** pane, as shown in the following screenshot:

Figure 8.21 – The Format tab

From the **Format** tab, you can edit various things.

11. Under **General**, you can configure which words should show up in the visual:

   - **Minimum number of repetitions**: How often a word has appeared in the data. If you set a minimum, only words that are used frequently will be included in the visual.

   - **Max number of words**: How many words (at most) you want to include in the visual.

12. When **Stop Words** is **On**, you can exclude stop words such as *the*, *a*, and *but*.

    There are two ways to exclude stop words:

    ▪ Use the default list created by Microsoft by switching the toggle for **Default Stop Words** to **On**.

    ▪ Use your own list by adding words in the field below **Words**. Separate the words with a space.

13. Set **Rotate Text** to **Off** to make sure all the words in **Word Cloud** are horizontal, instead of at an angle.

A Word Cloud offers you a way to visualize important words that have been found in a text dataset. It allows you to provide a quick overview of the many words that may be mentioned most often in your dataset, while also making the more commonly used words more salient.

Summary

In this chapter, we learned how to use the Language APIs offered by Azure Cognitive Services in Power BI. We looked at two approaches that give the same result. Both approaches allow you to extract key phrases, sentiment labels, or language from text. One approach is the simpler to use AI Insights feature, for which you need Power BI Premium. The other approach is to use Power Query Editor to make an API call, which gives you more flexibility but forces you to use a public privacy level on your dataset containing the text field. Whatever option you choose, you will end up with a new column containing insights that you can visualize using a Word Cloud.

In the next chapter, we will explore another NLU enrichment that we can add to our Power BI reports: adding question-answering with PowerApps to help users navigate the reports.

9

Integrating an Interactive Question and Answering App into Power BI

Visuals make a Power BI report. Your users will likely need more than graphical representations of the data to understand the complete story. Although you can add text to your reports, not all text should be put directly into the report.

End users of reports may have questions about the data, for example, to understand how a certain variable is measured or how often the data is refreshed. To avoid information overload in the Power BI report itself, you can integrate an app that will answer the frequently asked questions.

In this chapter, we'll be going step by step through creating an interactive question answering app, and integrating this with Power BI, in the following topics:

- Creating a question answering service
- Creating an FAQ app with Power Apps
- Integrating the FAQ app with Power BI
- Improving the question answering model

In this chapter, we'll use a **frequently asked questions (FAQ)** knowledge base as an example. As the steps we'll go through are dependent on each other, it is best to go through them in the order in which they are presented.

Technical requirements

There are four things you'll need to walk through the examples provided in this chapter:

- **Power BI Desktop**: You can install Power BI from the Microsoft Store, or find more advanced downloading options at `https://www.microsoft.com/en-us/download/details.aspx?id=58494`.

- **An Azure subscription**: Cognitive Services is a service available in Azure. To create a resource, you need to have access to an Azure subscription. If you don't have one yet, you can sign up for a free subscription at `https://azure.microsoft.com/en-us/free/`.

- **A license for Microsoft Power Apps**: To create an app with Power Apps, you need a work or school account and a license for Microsoft Power Apps, which may already be provided to you through your organization.

  If you want to use a personal email account, you can sign up for a 1-month free trial for Microsoft 365 for Business Basic here: `https://www.microsoft.com/microsoft-365/business/compare-all-microsoft-365-business-products`.

  To get a free license for Power Apps, select **Start free**, and sign in with your organizational account here: `https://powerapps.microsoft.com/`.

- **A glossary of terms**: To use the question answering service, we'll use a glossary of terms that already contains question-answer pairs. It is an overview of the terms included in the World Happiness Report data used in earlier chapters. Download the PDF here: `https://github.com/PacktPublishing/Artificial-Intelligence-with-Power-BI/blob/main/Chapter09/glossary%20of%20terms.pdf`.

Creating a question answering service

To create an interactive **question answering app** that is integrated into your Power BI report, we will use one of Azure's **Cognitive Services** offerings. One of the *Language APIs* of Cognitive Services offers an easy and quick way to create and manage a question answering service.

The question answering service is perfect for when we want to avoid a huge document with text that users have to navigate through. Instead, we want to offer a more intuitive way of getting the answer to a question. And what is more intuitive than simply being able to ask the question as you would to a colleague?

Understanding the application of question answering

What we are essentially creating in this chapter is a *smart search engine*. The information we want to share with our users is *static*. Think about a glossary of terms you may have for the variable names you included in your Power BI report, or more practical information you want to provide to users about the report itself. For example, users may wonder how often the data is refreshed, what the data sources are, or who to reach out to if they notice a mistake.

The static complementary information you want to share with the consumers of your Power BI reports may not be so crucial that they need to read it first, before even seeing the visuals. You probably want to offer this information just in case someone is looking for any part of that information. And, if they are looking for answers, you want to facilitate the search for them. Instead of having to search through many pages of a document, or maybe even multiple documents, you want to allow them to ask their question, and quickly get the most relevant answer in return.

Let's try to understand how the question answering service in Azure works with the help of *Figure 9.1*:

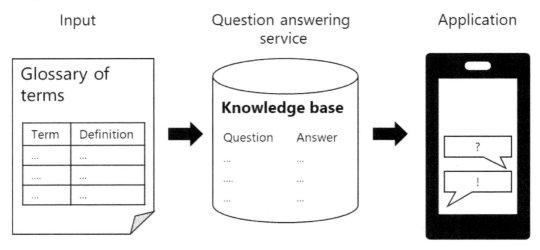

Figure 9.1 – Question answering flow

We start with the input on the left of the figure. The input to the service is whatever information you want to make more accessible through the smart search engine we will create. As an example, think about the glossary of terms we may include with a Power BI report. The glossary can contain all variable names that are used in the report and list them as terms, with a definition for each of these variables.

When the input document is provided to the question answering service, the service will convert each *term* to a **question**, and each *definition* to an **answer**, while making sure that each term/question is still paired with the appropriate definition/answer.

The question-answer pairs generated from the input are stored in the knowledge base in the question answering service. Once a knowledge base is *deployed*, it is available by making an *HTTP request* to the *URL* provided by the question answering service.

When a user asks a question in the app that we will create, we can ask the app to send an HTTP request to the knowledge base URL. The question that is asked will then be compared to the existing questions in the knowledge base, and the most similar question will be linked. The paired answer will be returned to the app and can then be presented to the user.

The purpose of using the question answering service is that we can search through our knowledge base of question-answer pairs with a question. In other words, the search query will be the question asked by the user. The service uses **Natural Language Processing (NLP)** to match the asked question to any of the existing questions provided in the knowledge base.

> **Keep It Simple**
>
> When adding questions to the question answering service, keep the questions as simple as possible. A question can be one or more words, such as *Country name* or *Social support*. There is no need to phrase the question in the service as an actual question, such as *What is social support?* or *What is meant by country name?* Often, it is better not to phrase it as a question as it creates distractions for the model, which result in mismatches. Just make sure to enter the key phrases you expect users to include in their queries.

Now that we have some grasp of what the question answering service does, let's learn more by *doing*. Much more will become clear when we experience the mechanics of the service and how it integrates with an application.

Configuring a question answering service

In the following steps, we will create a Cognitive Services resource and configure the question answering service for the FAQ app we want to integrate with Power BI later in this chapter.

You will need an email account that is associated with an Azure subscription to complete these steps. Make sure you have downloaded and saved the glossary of terms PDF on your local device from the following URL: `https://github.com/PacktPublishing/Artificial-Intelligence-with-Power-BI/blob/main/Chapter09/glossary%20of%20terms.pdf`:

1. Open a browser, and navigate to `https://language.azure.com/`.

2. Sign in with an account that has access to an Azure subscription.

3. A popup will appear and ask you to choose a language resource. We don't have one yet, so select **Create a new language resource**.

4. Select your **Azure subscription** version.

5. Set **Azure resource group** if you have an existing one (from previous chapters), or create a new one.

6. Fill in **Azure resource name**.

7. Select the location closest to you.

8. Select **F0** under **Pricing Tier**. Pricing tiers define the price and capacity of the service. **F0** is the free tier that is offered once per subscription and is ideal for experimentation.

    The result should look similar to *Figure 9.2*:

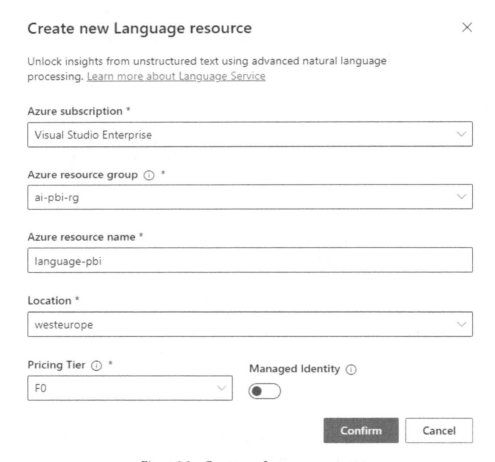

Figure 9.2 – Create new Language resource

9. Select **Confirm** and then **Done** to create the resource. The creation of the resource may take a couple of minutes.

    Once the resource is created, **Welcome to the Language Studio** appears.

10. Select **Answer questions**. This will take you down the page, as shown in *Figure 9.3*:

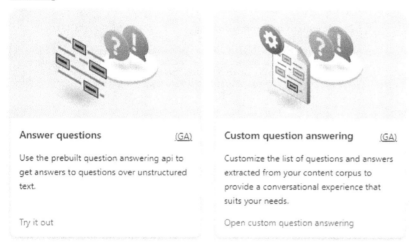

Figure 9.3 – Answer questions

11. Select the **Custom question answering** card.

12. On the **Custom question answering** page, as shown in *Figure 9.4*, select **+ Create new project**.

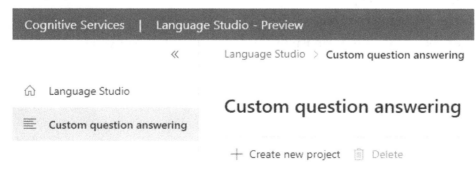

Figure 9.4 – Create new project

13. A popup appears. Select **Connect to Azure search**; this will open a new tab in your browser.

14. In the new tab, as shown in *Figure 9.5*, check the **Enable custom question answering** box.

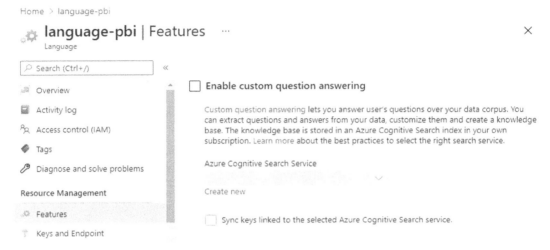

Figure 9.5 – Enable custom question answering

15. Under **Azure Cognitive Search Service**, select **Create new**. You'll be redirected to create another Azure resource that powers the search engine that makes the question answering possible.

16. Select an option for **Subscription** and **Resource Group**.

17. Enter a unique name.

18. Select the **Location** option closest to you.

19. Change **Pricing tier** to **Free**.

20. When you see something similar to *Figure 9.6*, select **Review + create**.

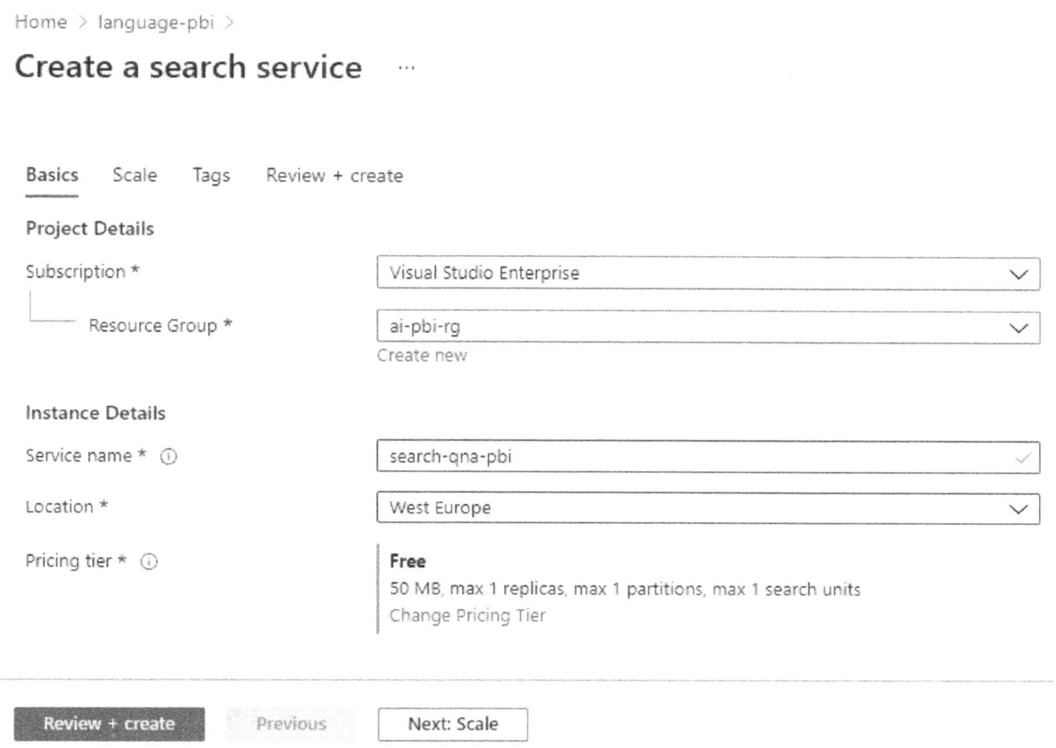

Home > language-pbi >

Create a search service ...

Basics   Scale   Tags   Review + create

Project Details

Subscription * | Visual Studio Enterprise   ⌄ |

└──── Resource Group * | ai-pbi-rg   ⌄ |
                            Create new

Instance Details

Service name * ⓘ | search-qna-pbi   ✓ |

Location * | West Europe   ⌄ |

Pricing tier * ⓘ | **Free**
                          | 50 MB, max 1 replicas, max 1 partitions, max 1 search units
                          | Change Pricing Tier

[Review + create]   Previous   [Next: Scale]

Figure 9.6 – Create a search service

21. After revising the summary of the search service you're about to create, select **Review + create** and then select **Create**.

22. Back in the pane where you can see **Enable custom question answering**, select the newly created **Azure Cognitive Search Service**, and select **Apply**.

23. Go back to **Language Studio**. You probably need to *refresh* the page to make sure it knows Azure Search has been connected.

24. Select the **+ Create new project** button again.

25. Set the language to **English** and select **Next**.

26. Enter your project name and select **Next**.

27. Select **Create project**.

A new project is created. You can add data sources, such as PDFs and Word documents with question-answer pairs. You can also add and edit the pairs directly in the Language Studio, with or without using a data source.

28. In the left menu, select **Manage sources**.

29. Select **+ Add source**. From the drop-down menu, select **Files** (see *Figure 9.7*).

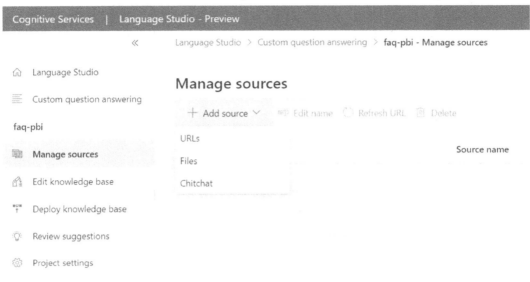

Figure 9.7 – Add data source

30. In the popup to add files, select the **glossary of terms.pdf** source you downloaded, and name it `glossary of terms`.

31. Select **Add all**.

    Once the source has been added, you can select the name to view and edit the question-answer pairs that are extracted (see *Figure 9.8*).

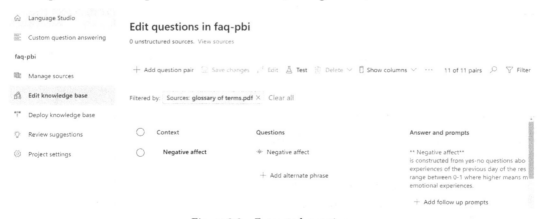

Figure 9.8 – Extracted questions

32. Select **Save changes** if it's enabled.

33. Select **Test** and a pane will open on the right.

In the pane, you can interact with the knowledge base by asking a question and receiving the best-linked answer. If you feel satisfied with the experience, you can continue and deploy the knowledge base. If you notice some questions are not answered accurately, you can go back and edit the knowledge base and test it again.

34. Select **Deploy knowledge base** from the left menu.

35. Select **Deploy**. A popup will appear to ask you whether you want to deploy this project.

36. In the popup, select **Deploy**.

Deployment can take a couple of minutes and will depend on how many question-answer pairs you have in your knowledge base. Once it is deployed, you will get a notification. The page for deploying your knowledge base now shows you information on when you last deployed it. To integrate the project with another application, you will need the prediction URL.

37. Select **Get prediction URL**. A popup will appear, as shown in *Figure 9.9*:

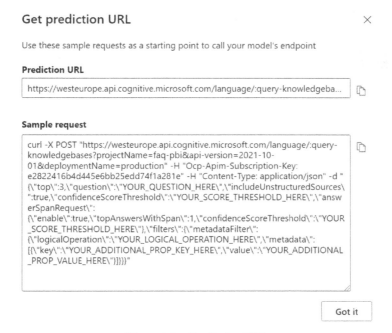

Figure 9.9 – Prediction URL

38. Copy the **Prediction URL** link and save it somewhere. We'll use it later when we'll call the service with Power Automate.

The prediction URL allows us to integrate the question answering service with any other application that can make an HTTP request to the URL. A sample request is provided below the URL, as shown in *Figure 9.9*. If you try and understand the URL and request, you'll notice that we need the following things whenever we make an HTTP request:

- The *URL* we need to send the request to
- The *key* to authenticate the request
- The *project name and details* of the knowledge base you want to retrieve an answer from
- The *question* you want to match to a question-answer pair from the knowledge base

Optionally, you can add more parameters to configure the result you get when making the HTTP request. For example, you can choose to return the top three matched answers instead of only the best-matched answer. For more information, check the following API reference: `https://docs.microsoft.com/rest/api/cognitiveservices/questionanswering/question-answering/get-answers`.

In the next section, we will use this prediction URL to integrate the question answering service with an application we create with Power Apps.

Creating an FAQ app with Power Apps

By creating the question answering service, we configured the model that will search for the best-matched question-answer pair. To interact with the service, we need a frontend that will allow the user to enter a question and see the response.

Power BI is part of the Microsoft Power Platform. Within this platform, we also have Power Apps and Power Automate. These two services allow us to quickly create the frontend we need.

To follow along with these steps, you need a work or school account and a license to Power Apps, as described in the *Technical requirements* section at the beginning of this chapter. As the process may seem complicated, the steps have been subdivided to create clarity.

Creating a new app with Power Apps

We'll start by creating a new blank app in Power Apps:

1. Open a browser, and navigate to `https://powerapps.microsoft.com/`.

2. Sign in with the work or school account that you associated with a Power Apps license.

3. On the Power Apps home screen, as shown in *Figure 9.10*, select **Blank app** and then **Create** under **Blank canvas app**:

Build business apps, fast

Create apps that connect to your data and work across web and mobile. Learn about Power Apps

Start from data

| Dataverse | SharePoint | Excel Online | SQL Server | Other data sources |

Make your own app

| Canvas app from blank | Model-driven app from blank | Portal from blank |
| Canvas app | Model-driven app | Portal |

Figure 9.10 – Power Apps home screen

4. Enter your chosen app name.

5. For **Format,** select **Phone**.

6.   Select **Create**.

A new project is opened with a blank phone canvas in the middle, as shown in *Figure 9.11*:

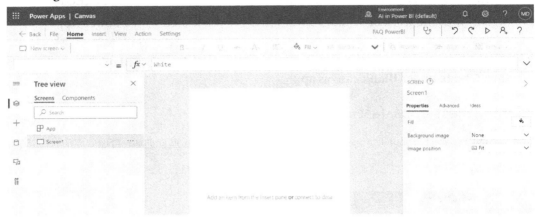

Figure 9.11 – Blank phone canvas

If you have never worked with Power Apps, this view might be overwhelming. There are many features visible on the current screen, which we won't use. So, let's just go through creating the app we need step-by-step, and you'll learn how to work with Power Apps along the way.

7.   As there is no autosave function in Power Apps, manually saving your work is very important. Select **File** from the top menu.

8.   Select **Save as**.

9.   The default option is to save the Power Apps in the cloud under the name you provided when creating it. Select **Save**.

10.   Go back to your app by selecting the left-pointing arrow at the top left of the screen.

11.   Once back in the editor, let's start by adding some content to our app. Select **Text** from the top menu. From the drop-down menu, select **Label**, as shown in *Figure 9.12*:

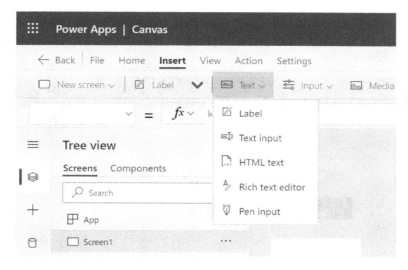

Figure 9.12 – Add label

A box with `Text` in it is now added to your app. Whenever you select the box, the menu on the right of your screen will show all details for the label. Feel free to play around with these options to explore how you can configure the label.

12. Edit the **Text** field in the menu to `Ask a question`, as shown in *Figure 9.13*. This will serve as the title of the app and will show our users what to do with it.

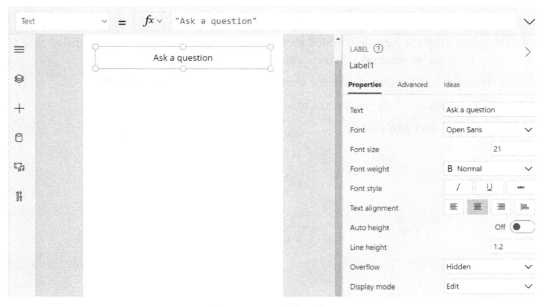

Figure 9.13 – Edit title

13. To ensure our users can type in their questions, let's add a *text input field*. Select **Input** from the top menu and select **Text input** from the drop-down menu, as shown in *Figure 9.14*:

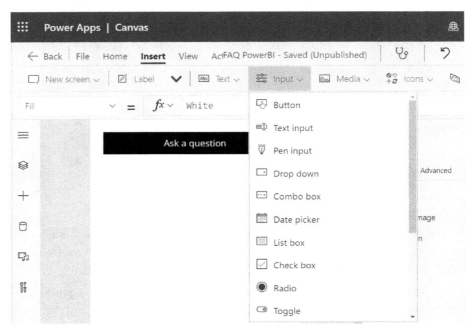

Figure 9.14 – Adding a Text input field

14. Drag and drop the new field added to your app to be centered under your title. Make it bigger so your users have room to type their questions.

15. In the right-hand menu, delete the words Text input from **Default**.

16. For **Hint text**, enter Type your question here.

17. For **Mode**, select **Multiline**.

The result for this input field should look like *Figure 9.15*:

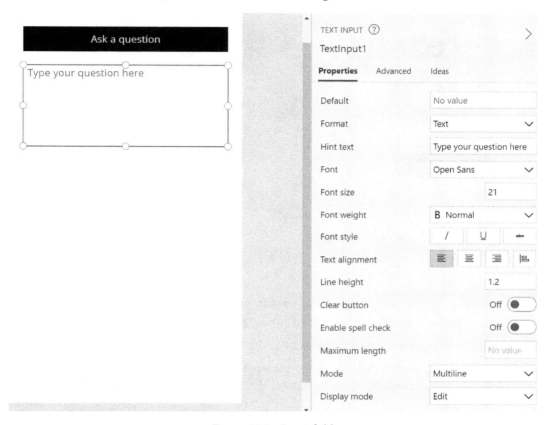

Figure 9.15 – Input field

18. From the top menu, select **Button**.

19. Drag and drop the new button field and center it below the text input field.

20. In the right-hand menu, as shown in *Figure 9.16*, change **Text** from Button to Get answer.

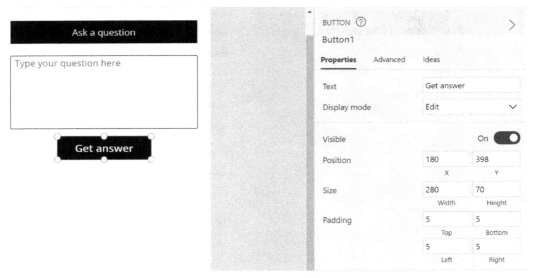

Figure 9.16 – Get answer button

Finally, we'll add another label that will show the answer to the question.

21. From the top menu, select **Text**. From the drop-down menu, select **Label**.

22. Drag and drop the text field and center it below the button, as shown in *Figure 9.17*. We'll change the content of the field later, after connecting the app with the question answering service.

Figure 9.17 – Answer field

23. Save your app.

The app now has all the content we need to make it work. We're covering the bare minimum here to make the app functional. Feel free to add anything you think will benefit the usability of the app when it is integrated with Power BI. Next, we'll use Power Automate to make the necessary HTTP request to the question answering service.

Adding Power Automate to call the question answering service

To connect Power Apps with the question answering service, we will use Power Automate to set up a workflow. The workflow will include an HTTP request to send the user's question to the prediction URL of the question answering service and get a matched answer back.

From a technical perspective, what we are trying to accomplish is to trigger the workflow whenever a user selects the button in the Power Apps app to get an answer. The workflow should take the content of the input text field and include that in the body of the HTTP request. The output of the workflow should be the matched answer, which then should be visualized in the Power Apps app below the button.

Let's create a Power Automate workflow to connect to the prediction URL of the question answering service we configured earlier in this chapter:

1. In Power Apps, select the **Get answer** button that you created in your app.
2. From the top menu, select **Action**.
3. From the options under the **Action** tab, select **Power Automate**.
4. A new pane appears as shown in *Figure 9.18*:

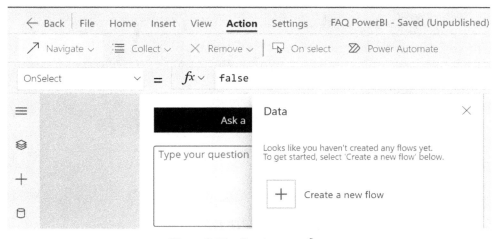

Figure 9.18 – Create a new flow

5. In the **Data** pane that just appeared, select **+ Create a new flow**.

6. A new tab will open in your browser to work with Power Automate while your app will remain open.

7. In the **Templates** overview, search for power apps.

8. Select the **Power Apps button** template. This will open a new workflow as shown in *Figure 9.19*:

Figure 9.19 – Power Apps trigger

The workflow will trigger whenever a user selects a Power Apps button. Let's add the next steps in the workflow.

9. Select **+ New step**.

10. Search for HTTP and select the **HTTP** search result to add it to your workflow. As a result, you should see a similar overview as *Figure 9.20*:

Figure 9.20 – HTTP action

11. For **Method**, select **POST**.

12. For **URI**, copy the *prediction URL* from the question answering service, and paste it here.

13. For **Headers**, add `Content-Type` on the left. In the same row, add `application/json`.

14. In the second row for **Headers**, add `Ocp-Apim-Subscription-Key` on the left, and add the *subscription key of the Cognitive Service* on the right. You can find the subscription key in the Azure portal, or in the sample request in the question answering service.

    For **Body**, we will send a key-value pair. The value will be the question that users enter in the app.

15. For **Body**, enter the following:

```
{
    "question": ""
}
```

16. Make sure your cursor is in between the last quotation marks. A popup should appear to help you add dynamic content. From the popup, select **Ask in PowerApps**. The result should look like *Figure 9.21*:

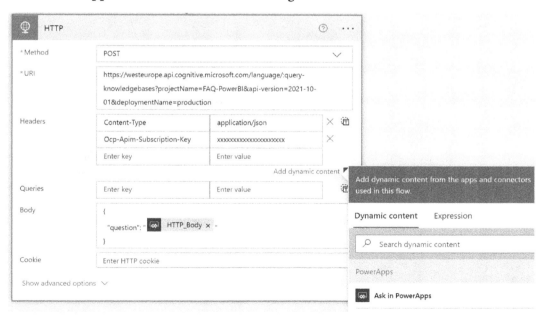

Figure 9.21 – Completed HTTP action

17. Add a new step to your workflow.

18. Search for `compose` and add the **Compose** operation to your workflow.

19. In the **Compose** step, for **Inputs**, type the following:

```
@{first(body('HTTP')?['answers'])?['answer']}
```

After you type in the expression, you should have the same overview as shown in *Figure 9.22*:

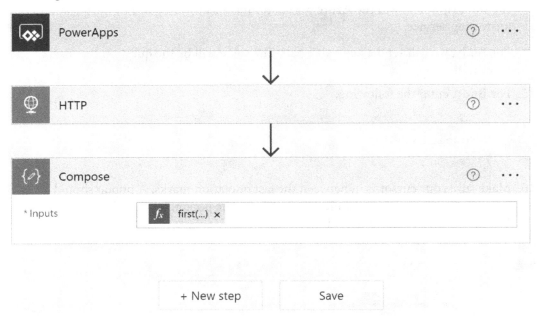

Figure 9.22 – Completed Compose action

Since the HTTP request assumes it may get multiple answers as a return, we need to specify that, from the HTTP output, we want to get the first result of the answers. In other words, we want the best-matched answer. By using the **Compose** operation, the best-matched answer is now a variable that we can use in our final step.

20. Add a new step to your workflow.

21. Search for `powerapp` and add the **Respond to a PowerApp or flow** action.

22. Select **+ Add an output**.

An overview of the different types of output is revealed, as shown in *Figure 9.23*:

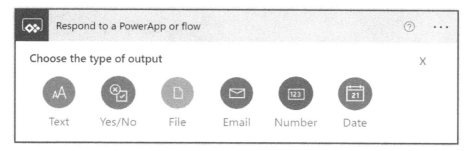

Figure 9.23 – Choose the type of output

23. Select **Text**.

24. For the title, enter generatedanswer. We will refer to this variable name in Power Apps, so make sure to enter the name correctly.

25. Select the box to enter a value to respond. The popup for dynamic content should appear, as shown in *Figure 9.24*:

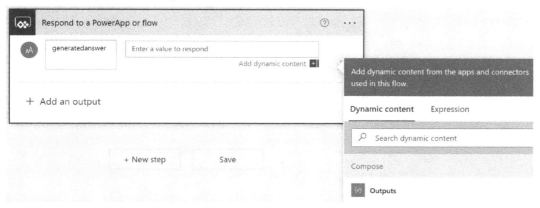

Figure 9.24 – Create a generatedanswer variable

26. From the dynamic content popup, select **Outputs**. This means we'll use the output of the **Compose** action, which is the best-matched answer that was returned by the HTTP request.

As a result, your workflow should look like *Figure 9.25*:

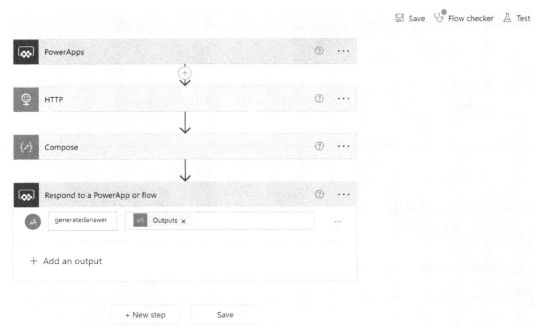

Figure 9.25 – Completed workflow

27. At the top right, select **Save**.

28. At the top right, select **Test**. This will open a new pane on the right, as shown in *Figure 9.26*:

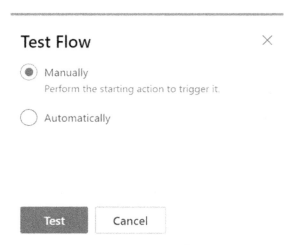

Figure 9.26 – Test flow

29. In the **Test Flow** pane, select **Manually**.

30. Select **Test**.

The pane (see *Figure 9.27*) will now show that to run the flow, we need to provide an input. This is the input that the workflow would normally take from the app. To test it, we'll manually add a question.

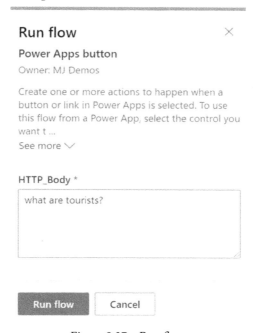

Run flow   ✕

Power Apps button

Owner: MJ Demos

Create one or more actions to happen when a button or link in Power Apps is selected. To use this flow from a Power App, select the control you want t ...

See more ∨

HTTP_Body *

what are tourists?

[Run flow] [Cancel]

Figure 9.27 – Run flow

31. Under **HTTP_Body**, enter what is GDP?.

32. Select **Run flow**.

You can view the completed workflow after it has finished running. Every time someone clicks the button in your Power Apps app, you will be able to view the executed workflow and check whether it has succeeded or failed. *Figure 9.28* shows what we get if we open up the last action and view the output that will be responded to the Power Apps app.

Figure 9.28 – Output workflow

The body of the outputs shows that a variable named `generatedanswer` holds the best-matched answer as its value. We will visualize this answer in our app in the next section.

We have created the workflow to take the input from our app and send it to the prediction URL of our question answering service. The answer is returned and is now ready to be visualized to complete our app.

Connecting Power Automate to Power Apps

We have created the necessary contents of the app in Power Apps, and the workflow to integrate with the question answering service in Power Automate. Now, we'll bring everything together and make sure the content of the text input field is taken as input for the workflow, as well as taking the output of the workflow and visualizing it in the text field in our app:

1. In Power Apps, select the **Power Apps button** option.
2. From the top menu, select **Action**.
3. From the **Action** tab, select **Power Automate** to open the **Data** pane on the right.
4. If there are no flows associated with **Button1** yet, as is the case in *Figure 9.29*, select **Power Apps button** from **Available flows**.

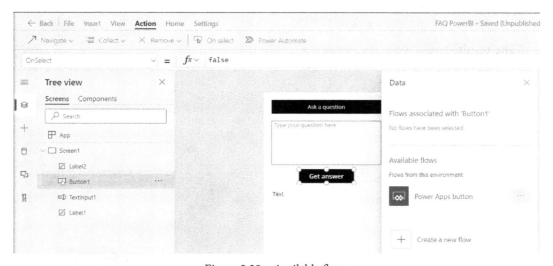

Figure 9.29 – Available flows

The Power Apps button workflow should now be associated with **Button1**, as shown in the **Data** pane in *Figure 9.30*:

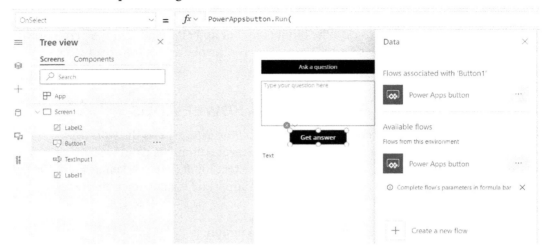

Figure 9.30 – Associated flow

You may also notice that in the formula bar at the top of *Figure 9.30*, we now see `PowerAppsbutton.Run`.

5.   Delete the current formula and enter the following:

```
Set(Response, PowerAppsbutton.Run(TextInput1.Text))
```

The result should look like *Figure 9.31*:

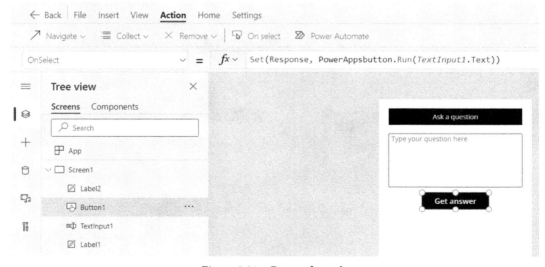

Figure 9.31 – Button formula

6. Select the label or text box below the button.

7. Replace the **Text** field with the following:

```
Response.generatedanswer
```

You should see something like *Figure 9.32*:

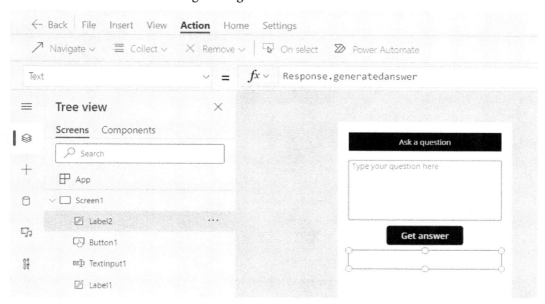

Figure 9.32 – Text formula

Our app is ready!

8. Test it by selecting the play button at the right top of the screen.

9. All editing options will disappear to simulate the user experience, as shown in *Figure 9.33*. Type a question and select the **Get answer** button to test it.

Figure 9.33 – Final app

Finally, save your work and publish the app so that we can integrate it with Power BI.

10. Select **File** from the top menu.

11. **Save** your work.

12. After saving, select **Publish** and confirm. Without publishing, the app will not be visible to anyone else within our organization.

We have created an app with Power Apps that is connected to the question answering service through Power Automate. The app can be shared as a standalone app within your organization. However, we created it to integrate it with Power BI, so let's do that next.

Integrating the FAQ app with Power BI

Finally, we want to make the FAQ app we created with Power Apps and the question answering service available to end users of our reports in Power BI. To do that, we'll add the app we created as a visual to our report:

1. Open a Power BI report in Power BI Desktop.
2. Sign in with the same email account you used for Power Apps and Power Automate.
3. From the **Visualizations** pane, select **Power Apps for Power BI**.

    To integrate with Power Apps, you need to add data to this visual. Since we are not relying on the Power BI data with our app, you can essentially use any data (simple mock data will do). Use any already loaded data if you want, otherwise, you can quickly create a field with the following three steps.

4. From the **Home** ribbon, select **Enter data**.
5. Create a table with one column and one row and name the table and column PowerApps, as shown in *Figure 9.34*:

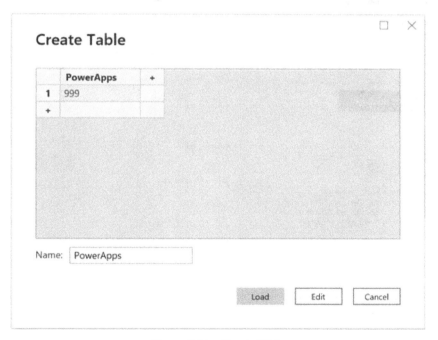

Figure 9.34 – Create Table

6. Select **Load**.

7. Ensure the Power Apps visual is selected and add a data field from your dataset to the **PowerApps Data** field.

   Once data is added to the Power Apps visual, the visual asks you to choose an existing app or create a new one.

8. In the Power Apps visual, select **Choose app**.

9. Select the app you created and select **Add**.

10. A browser may open again for authentication. Sign in again if necessary. You may also be prompted to save the app; save or skip the step to continue.

    As a result, the app will appear in your Power BI report, as shown in *Figure 9.35*:

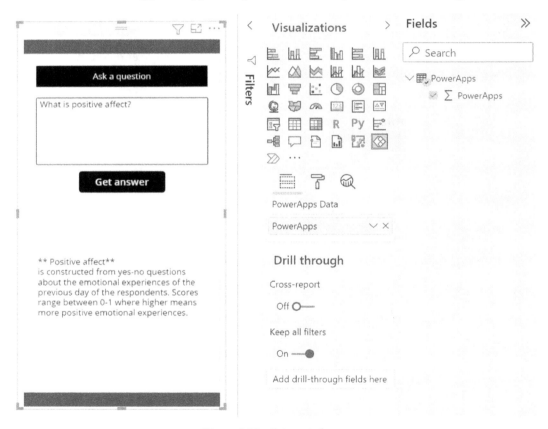

Figure 9.35 – Integrated app

The app we created in Power Apps is now integrated with Power BI. Once you publish your completed report to the Power BI service and share the report with others, users will be able to use the app to get answers to their questions.

As with many applications and AI models, we of course do need to monitor and keep improving over time. It is highly likely that some of our users' questions will remain unanswered because the question answering service can't find a good match. We can review the predictions of the question answering service and use them to improve the knowledge base model. To get an idea of how that would work, let's explore how we can optimize the question answering model in the next section.

Improving the question answering model

It is very important to monitor and continuously improve a machine learning model. The same goes for the model we created when using the question answering service. Even though it works perfectly when we as designers test it, there is no guarantee that it will stand the test of time.

Over time, new questions may emerge, or users may interact with your app differently. For example, users may ask shorter questions as they get used to working with the question answering service through the app.

There are two main things we can do to improve the model:

- **Use active learning in the Language Studio**: An easy approach that analyzes predictions for you. Whenever questions seem too similar, the service concludes you need to provide more information to disambiguate between questions more clearly. Suggestions are provided, and you can review them in the Language Studio in the **Review suggestions** tab.

- **Log diagnostics in Azure**: A more advanced approach to collecting telemetry data on your knowledge base. With diagnostics, you can monitor content such as unanswered questions. More information can be found here: `https://docs.microsoft.com/azure/cognitive-services/language-service/question-answering/how-to/analytics`.

> **The model doesn't automatically improve over time**
>
> There is no way for the knowledge base to automatically improve over time. This is by design. If your model would take all user interactions to improve the model, users can very easily sabotage your solution. If you want to improve the model, you have to review the changes before initializing retraining.

A best practice across many types of AI applications is to have a human in the loop to ensure a good and fair model. This is why, even with active learning, you are required to either accept or reject suggestions. You are free to create your own automated workflow, which you are then fully responsible for. We'll go further into the topic of responsibility in *Chapter 13, Responsible AI*.

Summary

In this chapter, we have explored the question answering service and other parts of the Power Platform. With the combination of these services, we were able to offer an intuitive app in a Power BI report to help users navigate through the data. Whether you want to provide them with an easy way to learn more about the data you included and how you collected it, or you want to give them more practical information about the report; instead of adding lengthy documents that are hard to navigate through and daunting to read, we have now learned how to create an app that can search through a knowledge base for us.

Up until this chapter, we have explored how we can work with text in Power BI. In the next chapter, we'll learn how we can work with images using computer vision.

10

Getting Insights from Images with Computer Vision

An image is worth a thousand words, but it can be hard to extract those words from an image. This expression shows how complex images are and how much information they can contain, which is why getting insights from images is a challenging task.

To quickly get insights from images, you can use **Computer Vision** models. **Computer Vision (CV)** is a type of **Deep Learning (DL)**, where the goal is to teach the computer to understand images. With the Computer Vision models offered by **Azure's Cognitive Services**, you can analyze images in Power BI.

In this chapter, we'll explore two ways to work with the Vision **application programming interfaces (APIs)** from Cognitive Services in Power BI: **Computer Vision** and **Custom Vision**. Additionally, we'll visualize a set of images in Power BI to add context to the extracted insights. Specifically, we'll be looking at the following topics:

- Getting insights with Computer Vision using AI Insights
- Configuring Custom Vision

- Integrating Computer Vision or Custom Vision with Power BI
- Using visuals to show a reel of images in a report

Let's go over some technical requirements before we start working with images.

Technical requirements

To walk through the examples provided in this chapter, you will need the following:

- **Power BI Desktop**

    You can install Power BI from the Microsoft Store or find more advanced downloading options here:

    `https://www.microsoft.com/en-us/download/details.aspx?id=58494`

- **An Azure subscription**

    Cognitive Services is a service available in Azure. To create a resource, you need to have access to an Azure subscription. If you don't have one yet, you can sign up for a free subscription here: `https://azure.microsoft.com/en-us/free/`.

- **Random images sample dataset**

    For the AI Insights example, we'll use a small dataset consisting of some miscellaneous objects. All these images were taken by the author. You can find the images here: `https://github.com/PacktPublishing/Artificial-Intelligence-with-Power-BI/tree/main/Chapter10/random-images`.

- **Clothing images sample dataset**

    For the Custom Vision examples, we'll use a clothing images dataset. As images require a lot of compute power, only a few images are provided in the sample dataset. The images are taken either by the author or from www.unsplash.com. You can find the images here: `https://github.com/PacktPublishing/Artificial-Intelligence-with-Power-BI/tree/main/Chapter10/clothing-images`.

With Power BI Desktop, an Azure subscription, and a set of images, we'll have extracted insights at the end of this chapter.

Getting insights with Computer Vision using AI Insights

The easiest way to analyze images in Power BI is by using the Computer Vision model provided to us through the **AI Insights** feature in Power BI. The AI Insights feature in Power BI does exactly the same as the Computer Vision API from Azure's Cognitive Services.

When we use a pretrained Computer Vision model such as the one offered by Cognitive Services and AI Insights, we don't have to provide any training dataset and can use the model to analyze our images.

Using a pretrained model brings with it advantages and disadvantages. As already discussed in *Chapter 7, Using Cognitive Services*, the purpose of these pretrained models is to *reduce the time and expertise* needed by someone such as you to apply them. However, pretrained models are often trained on a generic dataset, which makes them applicable to many different situations.

It may be of course that for your specific use case, using a pretrained model doesn't give you the information you need. For example, if you simply want to extract generic tags that can be associated with images to get an idea of what the images contain, using the Vision option of AI Insights or the Computer Vision API from Cognitive Services in Power BI is exactly what you need. Take, for example, objects such as houses, laptops, or apples you can detect from images.

If, on the other hand, you have a dataset and you want to know which item of clothing shows in a specific photo, you may need to train your own model. Thankfully, Cognitive Services' Custom Vision offers you the ability to work with a pretrained model to save time and expertise, as well as add your own training data to customize the model for your use case.

So, if you want to get quick insights from a set of images you are working with in Power BI, continue reading to learn more about using AI Insights' Vision option. In the next sections, *Configuring Custom Vision* and *Integrating Computer Vision or Custom Vision with Power BI*, we'll go into more detail on how and when to use Custom Vision.

Using the Vision option of AI Insights

To use the AI Insights Vision feature in Power BI, there are a couple of requirements your Power BI license and your data need to satisfy. Before we cover these requirements, let's review what AI Insights are (we also used it for language in *Chapter 8, Integrating Natural Language Understanding with Power BI*).

Azure's Cognitive Services include many pretrained models that we can use in any application, by calling the provided API. Using the Computer Vision API from Cognitive Services requires you to create the necessary resources in Azure and work with API calls. To skip all that work, you can use AI Insights instead to tag images.

The Computer Vision API of Cognitive Services offers more options in the information it can extract from your images, such as recognizing landmarks or celebrities. If you need more than image tags, the *Integrating Computer Vision or Custom Vision with Power BI* section in this chapter will guide you through how to do that.

If you choose to use AI Insights instead, you need to have a **Power BI Premium license**. Once you have access to that, using AI Insights is straightforward; you need to import a dataset that includes a column with either of the following:

- **Binary content** of images
- **Uniform Resource Locators** (**URLs**) of images

When you trigger AI Insights to tag your images, a new column will be added with the tags. Let's see it in action.

> **Random images used for walkthrough**
>
> For this walkthrough, we have provided some random images for you to store in a local folder after downloading them from here:
>
> ```
> https://github.com/PacktPublishing/Artificial-
> Intelligence-with-Power-BI/tree/main/Chapter10/
> random-images
> ```

We'll start by loading in data referring to images, after which we'll add the insights extracted from the images with AI Insights. Follow these steps:

1. Open a new Power BI report in Power BI Desktop.
2. In the top ribbon, in the **Home** tab, select **Get data** and then **More...**.
3. Select **Folder**.
4. Browse to the folder path of the folder in which the images (and no other files) are stored.
5. After reviewing the preview of the data, select **Transform Data**. The imported data will resemble this:

Figure 10.1 – Images in a table

The preceding table shows the metadata of the images, including the filename, extension, date created, and other things. Most importantly, there is a column called `Content` with a binary representation of the images. Feel free to remove any column except the `Content` column for this exercise. We won't visualize the images with this binary content in this example, but we will use it to extract insights.

6. In the top ribbon, in the **Home** tab, select **Vision** from **AI Insights**. A popup will appear, as shown in the following screenshot:

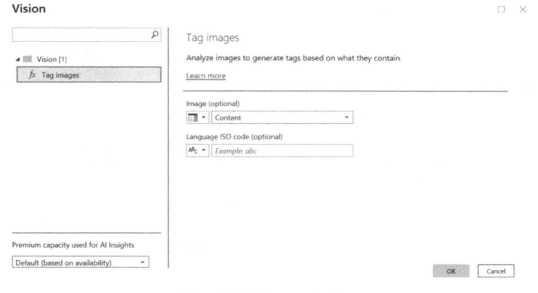

Figure 10.2 – Vision configuration

7. In the **Vision** pane, select **Tag images**.

8. For **Image (optional)**, select Content.

9. Select **OK**.

Each image will be sent to the Computer Vision model to get scored, which means that the analysis may take a while. The more images you have, the longer this process will take. Once the insights have been extracted, you'll see that rows have been duplicated, as shown in the following screenshot:

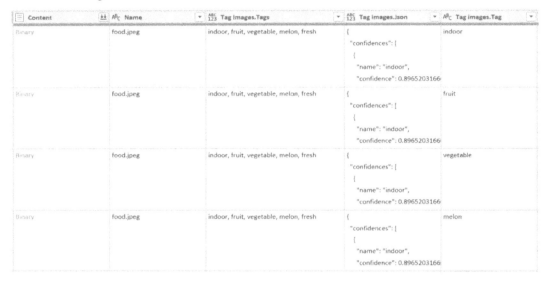

Figure 10.3 – Image tags

AI Insights adds the following columns after using **Vision**:

- `Tag images.Tags`: All tags extracted from the image.

- `Tag images.Json`: The complete result from the model, shown as **JavaScript Object Notation (JSON)**. Parts of the result, such as the tags and confidence score, are extracted to other columns.

- `Tag images.Tag`: One of the tags extracted from the image.

- `Tag images.Confidence`: The confidence score for `Tag images.Tag`. The value will range between 0 and 1 and indicates how certain the model is about the predicted tag. When the confidence is close to 0, the model is very unsure. The closer the score is to 1, the more certain the model is about the tag.

- `Tag images.ErrorMessage`: Hopefully, the value for this column is `Null`, indicating no error has occurred. If an error did occur, the message will show here. For example, if the folder also contained non-images such as text files, the message may say that the input data wasn't in the right format.

You can choose which column you want to use. For example, you can create a word cloud of the `Tag images.Tag` column, as shown in *Chapter 8, Integrating Natural Language Understanding with Power BI.*

Now that we have seen how to use AI Insights **Vision** to get tags from images in Power BI, let's explore another option that gives us more flexibility in terms of the information we want from the images: Custom Vision.

Configuring Custom Vision

CV is a form of **supervised learning** (**SL**) as it needs to have labeled data to train the model. A model will only be able to recognize tags or labels in images if it has seen them before. For example, if you train a model only on images of cats and dogs, it will not be able to identify what a hamster is.

In that sense, Computer Vision models learn in a similar way to children. They learn by examples, and the more examples they experience, the better they become at recognizing objects.

Now, what if we want to extract tags from images that are specific to our use case? For example, imagine you run a secondhand clothing store. Normally, customers would bring their clothes to the store, and you decide whether they are worth something and whether you want to sell them in your shop. Especially with the pandemic, you may want to avoid customers coming to your shop for clothes you don't want to accept anyway.

Instead of customers bringing clothes to the shop, you ask them to take photos of the items of clothing and send them to you. That way, you can browse through the images to determine whether you want to add items to your collection.

In this case, using a generic pretrained model may tell us the image contains clothing, but not what category of clothing. To create a more specialized model, you can use Custom Vision. With Custom Vision, you use a partially pretrained model to avoid having to start from scratch. You then add examples of labels you want to extract from future photos to create a model catered to your needs.

In the example, we want a model that recognizes whether an item of clothing falls into any of the categories of clothing—such as shirts, pants, skirts, and dresses—that you sell in the shop. By automatically categorizing the photos, it is easier to get insights into what customers are trying to turn in. If we have a shortage of dresses, for example, we'll just focus on photos of dresses that have been sent.

Preparing the data for Custom Vision

Let's create a model with Custom Vision that categorizes images of clothing. To download the `Clothing` folder with images, go to the GitHub repository at `https://github.com/PacktPublishing/Artificial-Intelligence-with-Power-BI/tree/main/Chapter10/clothing-images`.

The `Clothing` folder contains three folders: `dress`, `jacket`, and `tshirt`. Each folder contains 9 or 10 images showing the relevant item of clothing. In this example, we use a very small set of images to focus on showcasing the features of Custom Vision, while minimizing the training and scoring time.

In real-life applications, there are some data requirements you need to consider, as outlined here:

- **Data quantity**: Custom Vision recommends at least 50 images per category. DL models commonly improve in accuracy if more data is used for training. Remember that Custom Vision uses a partially pretrained model, meaning that the training dataset you add doesn't need to be as large as when you are training a Computer Vision model from scratch.

- **Data quality**: Focus on safeguarding the data quality when adding more images. What is considered good quality depends on your use case. Remember that models only know what they are taught, so if you only teach a model to recognize a can of soda when it is centralized in a photo, showing the logo, the model may not recognize the can when only part of it is shown on the edge of an image. Data can be manipulated to ensure you have a diverse **variety** of images.

- **Balanced dataset**: When training a model for image classification, you may have an imbalanced dataset—for example, 90% of images show jackets, while only 10% of images show a dress. An imbalanced dataset may reflect reality, as the shop also holds more jackets than dresses. However, if you train your model on an imbalanced dataset, your model may become biased and assume any image should be tagged as a jacket. Evaluation metrics (as we will discuss after training the model) help in recognizing bias. To reduce bias, retrain your model after including more images of the underrepresented category in your training dataset. You can either collect or create more images. Create more images by reshaping and enhancing existing images (to make them look different from the model), or by using a method such as **Synthetic Minority Oversampling TEchnique (SMOTE)**.

- **Other category**: If an image doesn't resemble any of the categories included during training, the model is likely to not classify it as any label. However, if the image resembles your categories but still needs to be excluded, an **Other** category can help. For example, you may receive images with sports clothes, which resemble the clothing labels you included in your training data. To avoid the model from classifying the sports clothes as a normal T-shirt or jacket that you want to sell, you want the sports clothes to be categorized as **Other** by the model.

Custom Vision allows you to quickly train a model to detect your own created labels. It needs very few images to already create an adequately performing model, as we will see in the next section. Even though the dataset used in the examples is too small according to the data requirements listed previously, it will minimize training and scoring time and serve its purpose for now. Just remember that in real-life applications, you want to consider the data quantity, quality, balance, and the **Other** category when training a model in Custom Vision.

Training the model in Custom Vision

Training a model is an iterative process; the same goes for when you train a model in Custom Vision. The benefit of using Custom Vision is how little data you need and how quickly you can train a model. In the previous section, we went over some data requirements. Although these requirements indicate how we can improve the model's performance, don't worry about getting the perfect dataset when you start. It is best to start experimenting with any data you have, evaluate the model, and explore how you can improve it by tweaking the training data.

Let's train a model in Custom Vision with the small dataset of items of clothing we have to learn about the process.

You need a Cognitive Services resource to use Custom Vision. To create one, go to *Chapter 7*, *Using Cognitive Services*, and walk through the *Creating a Cognitive Services resource* section. Next, follow these steps:

1. Navigate to `www.customvision.ai`.
2. Sign in with the account that is associated with the Azure subscription.

Once signed in, you should see a **Projects** overview, as shown in the following screenshot:

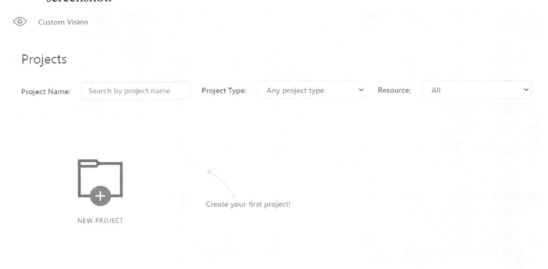

Figure 10.4 – Projects overview

3. Create a new project by clicking on **NEW PROJECT**.

4. For **Project Name**, enter Clothes categories.

5. Choose the Cognitive Services resource you created.

6. For **Project Type**, select **Classification**.

7. Select **Multiclass**.

    As customers will upload photos of one item of clothing, you'll expect to only get one category per image. If multiple tags can be created for an image, use **Multilabel**. If you expect multiple objects in an image, use **Object Detection** as the **Project Type** value.

8. For **Domains**, select **Retail**.

9. Select **Create Project**.

    Once our project is created, we will see the project page, as shown in the following screenshot:

Figure 10.5 – Project page

10. Select **Add images**.

11. Browse to the `dress` folder you downloaded and select **All images**.

12. Under **My Tags**, type `dress` and press *Enter*.

13. Before you upload the files, verify whether the **Image upload** pane resembles the one shown here:

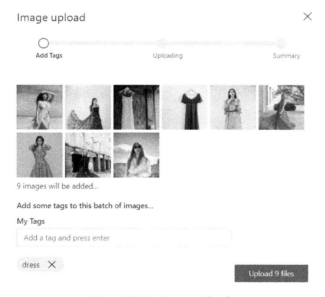

Figure 10.6 – Image upload

14. After uploading the 9 images with dresses, repeat *Steps 9*, *10*, and *11* for `jacket` and `tshirt`.

15. Once all images are uploaded select **Train**.

16. Opt for **Quick Training**.

Normally, training a DL model that classifies images can take hours, if not days. *Quick training* means we can do it within minutes. The more training data you have, the longer training will take. Since we have so little data, expect training time to be no longer than 30 minutes.

When the model is finished training, you can view the evaluation metrics of the model in the **Performance** tab, as shown in the following screenshot:

Figure 10.7 – Model performance

No model is the same, so you may get different evaluation metrics; especially with so little data, there can be a lot of variance between models. Before we can fully understand what we see in the **Performance** overview, let's go over some key concepts for evaluating classification models.

Evaluating classification models

Whenever we work with classification models, one plot that can help evaluate the model is a confusion matrix. To create a confusion matrix, the model takes a set of images (of which the tag is known) and tries to predict for each of those images what the tag should be. As a result, we can compare the actual tags with the predicted tags to evaluate the model.

Let's take an example from our data and start with the `dress` tag. Based on the images tagged as dresses, and the model's predicted images that are dresses, the following metrics are calculated:

- **True Positive** (**TP**): The number of images that actually show a dress, *and* for which the model predicted there to be a dress.

- **True Negative** (**TN**): The number of images that do not show a dress, *and* for which the model predicted there to be no dress.

- **False Positive** (**FP**): The number of images that do not show a dress, *but* for which the model predicted there to be a dress.

- **False Negative** (**FN**): The number of images that do show a dress, *but* for which the model predicted there to not be a dress.

These four metrics together make the structure of a confusion matrix, as shown in the following screenshot:

	Actually dress	**Actually not dress**
Predicted dress	True Positive	False Positive
Predicted not dress	False Negative	True Negative

Figure 10.8 – Confusion matrix

To evaluate a classification model, the confusion matrix will show the number of images per metric. Based on the requirements of the model and the evaluation metrics, you may decide to retrain your model after adding more data of certain categories to improve it.

Now, back to the performance metrics shown in *Figure 10.7*. The three evaluation metrics that are shown are outlined here:

- **Precision**, calculated as TP/(TP+FP). Of all images actually showing a dress, how many were predicted to have a dress?

- **Recall**, calculated as TP/(TP+FN). Of all images predicted to have a dress, how many actually showed a dress?

- **A.P.:** If we plot the **precision-recall (PR)** curve, the **area under the curve (AUC)** is the **average precision (AP)**. Used to balance between precision and recall to avoid over- or underfitting.

These three metrics are often shown as percentages where closer to 100% indicates a more accurate model. However, interpreting the results should always be done with care. For example, in the case of clothing, we want the model to perform well over all categories.

In other cases, we may want the model to perform better in one category, at the cost of another. Let's say you are working for a city that only allows electric cars. At the borders of the city, you may install cameras that take images of cars and classify them as electric or not. You want to catch the non-electric cars as they should get a fine for entering the city. In this case, you want the model to find all non-electric cars, even if that means that sometimes, an electric car is incorrectly tagged as non-electric. The city may prefer those people to object and get their fine waivered instead of missing cars that should be fined, for the sake of the air quality in the city.

In other words, evaluating a model can be challenging, which is why it is often up to the data scientist. A good rule of thumb is to aim for higher percentages across all evaluation metrics, although each use case has different requirements.

One final thing to highlight about Custom Vision's **Performance** overview is the breakdown of the evaluation metrics per category, as shown at the bottom of *Figure 10.7*. Next to the metrics, Custom Vision shows the **Image count** value per category. This will help you to detect whether you have an imbalanced dataset that may skew the model or not enough images in general. Custom Vision advises using at least 50 images per category.

If you notice that you don't have enough images of a certain category, there are some techniques to *impute* data. One approach is to use the images you have but slightly alter them to create more images—for example, you can cut part of an image to create a new image, you can rotate an image, or you can change the coloring of an image.

As always, creating a model is an iterative process; so, experimenting with different datasets and images and retraining your model is expected before landing on a model that is acceptable for your use case.

Once it is acceptable, we want to use the model. To consume the model from any other application, we need to have an endpoint and key. We'll create that next.

Publishing your Custom Vision model

When you have an iteration of a trained model in Custom Vision that you want to use in an application, you publish the iteration. Once the iteration is published, you can use the generated endpoint and key to make an API call to the model and use it to tag new images.

In the following steps, we'll publish the Custom Vision model to get the endpoint with which we can integrate it with Power BI:

1. In the **Performance** tab, select **Publish**.

2. For **Model name**, enter `clothes-classifier`.

3. For **Prediction resource**, select the Cognitive Services resource you created.

4. Select **Publish**.

5. Back in the **Performance** overview, select **Prediction URL**. You'll see two options for how to use the **Prediction** API, as shown in the following screenshot:

Figure 10.9 – Prediction URL

Depending on whether you use an image URL or the file (or binary content) of an image, you can use a different endpoint. In the *Integrating Computer Vision or Custom Vision with Power BI* section, we'll use an image URL.

6. Copy and save the endpoint under **If you have an image URL**.

7. Copy and save the value of your prediction key.

Whenever you want to classify images and tag them with your own labels, you can use Custom Vision. We have learned how to use Custom Vision, how to configure it, and how to publish it. Next, we will integrate the clothing classifier model with Power BI.

Integrating Computer Vision or Custom Vision with Power BI

Whether you want to integrate a Computer Vision or Custom Vision model from Cognitive Services with Power BI, the process is similar. With both Azure Cognitive Services, you can use the model to score new data by using the endpoint and one of the authentication keys that will be generated.

Using the endpoint and key of a model, you can also integrate a Computer Vision or Custom Vision model in Power BI. The API call to the service is similar to the way we integrated language with Power BI in *Chapter 8*, *Integrating Natural Language Understanding with Power BI*.

Let's start by importing data in Power BI that includes image URLs. You can download the metadata table as a **comma-separated values** (**CSV**) file from GitHub at `https://github.com/PacktPublishing/Artificial-Intelligence-with-Power-BI/tree/main/Chapter10/image-urls`. Then, proceed as follows:

1. Open the **Power BI Desktop** application.

2. Add a new **Data Source** value.

3. Select **Text/CSV**.

4. Select the `image-urls.csv` metadata table and load the data.

   A table called `image-urls` will be created, containing one `Folder Path` column that contains the image URLs, as shown in the following screenshot:

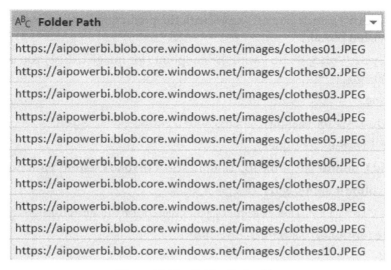

Figure 10.10 – Folder Path table

5. In **Power Query Editor**, add a **New Source**.

6. Select **Blank Query**.

7. Select the new query and select **Advanced Editor**.

8. Delete all current content, and paste in the following code:

```
// Returns image tags
(image) => let
    apikey      = "<your-prediction-key",
    endpoint    = "<your-endpoint>",
    jsontext    = Text.FromBinary(Json.FromValue(Text.
Start(Text.Trim(image), 5000))),
    jsonbody    = "{ Url: " & jsontext & " }",
    bytesbody   = Text.ToBinary(jsonbody),
    headers     = [#"Prediction-Key" = apikey, #"Content-
Type" = "application/json"],
    bytesresp   = Web.Contents(endpoint,
[Headers=headers, Content=bytesbody]),
    jsonresp    = Json.Document(bytesresp),
    tags  = jsonresp[predictions]{0}[tagName]
in  tags
```

9. Replace `<your-prediction-key>` with the prediction key you copied from the Custom Vision portal.

10. Replace `<your-endpoint>` with the endpoint (for image URLs!) you copied from Custom Vision.

11. Select **Done**.

12. Rename `Query1` for usability. Change it to `Classify images`.

13. Select the `images-url` query.

14. In the **Add Column** tab of the top ribbon, select **Invoke Custom Function**.

15. For **New column name**, enter `Image tags`.

16. Select the `Classify images` query you created for the **Function query** field.

17. For **image (optional)**, select `Folder Path`.

18. Select **OK**.

The result is that a new `Image Tag` column will have been added, as shown in the following screenshot:

A^BC Folder Path	ABC 123 Image Tag
https://aipowerbi.blob.core.windows.net/images/clothes01.JPEG	jacket
https://aipowerbi.blob.core.windows.net/images/clothes02.JPEG	jacket
https://aipowerbi.blob.core.windows.net/images/clothes03.JPEG	jacket
https://aipowerbi.blob.core.windows.net/images/clothes04.JPEG	jacket
https://aipowerbi.blob.core.windows.net/images/clothes05.JPEG	t-shirt
https://aipowerbi.blob.core.windows.net/images/clothes06.JPEG	jacket
https://aipowerbi.blob.core.windows.net/images/clothes07.JPEG	t-shirt
https://aipowerbi.blob.core.windows.net/images/clothes08.JPEG	dress
https://aipowerbi.blob.core.windows.net/images/clothes09.JPEG	t-shirt
https://aipowerbi.blob.core.windows.net/images/clothes10.JPEG	jacket

Figure 10.11 – Image Tag column

For each image, we have now added the tag that the Custom Vision model detected. With this information, we can visualize the images in Power BI, including the insights extracted from them. In the next section, we'll see how we can bring images and insights together.

Using visuals to show a reel of images in a report

Although you may want to analyze images to visualize only the insights extracted in your Power BI report, it is also possible to show a set of images.

In the example used throughout this chapter, we used images of items of clothing that we wanted to categorize. After training a model in Custom Vision, we were able to score the images in Power BI and get the category of clothing detected in each image.

In this section, we'll explore how we can visualize a reel of images.

Storing data and ensuring it is anonymously accessible

To show images in Power BI, they need to be anonymously accessible. Whether you want to refer to the image URL or the binary contents you have stored somewhere, either should be publicly accessible without requiring authentication. That means you can store it on OneDrive if the folder is publicly accessible, on a public website of some kind, or—for example—on a cloud storage solution such as Azure Blob storage.

To store images on a Blob storage, you'll need to create an Azure Blob storage account using an Azure subscription. Blob storage is the cheapest storage for unstructured data. You store images in a so-called container within the Blob storage. The container can be made public so that no authentication is required to access the data in the container.

In the clothes example, a Blob storage instance has been created for you, and you only need the URL for each image to retrieve the images and visualize them in your report. Feel free to create your own Azure Blob storage and copy the images to your own container. You can download the images from GitHub as well at the following link: `https://github.com/PacktPublishing/Artificial-Intelligence-with-Power-BI/tree/main/Chapter10/clothes-images/untagged`.

Assuming you imported the `image-urls` dataset in the previous section, and therefore are using the Blob storage that has already been created, you can visualize the images with the following steps:

1. Ensure **Power Query Editor** is closed and all changes to your data have been applied.
2. Change the **Data Category** value of `Folder Path` to `Image URL`.
3. In the **Visualizations** pane, select a matrix.
4. From the `image-urls` table, select `Folder Path` and drag it to **Rows** if you want to see the images presented vertically, and to **Columns** if you want to see the images presented horizontally.

For example, the following screenshot shows the images horizontally:

Figure 10.12 – Visualized images

5.   To make the images larger, go to **Format**, and under **Visual**, increase the **Image height** value.

Next to using a matrix, you can also visualize images using a table, a slicer, or a multi-row card. The advantage of using a matrix is that you can choose whether you want to show the images horizontally or vertically.

When adding another visualization such as a stacked column chart showing the number of images per category, as shown in the following screenshot, you can use the extracted insights to learn more about the contents' images in a more intuitive overview:

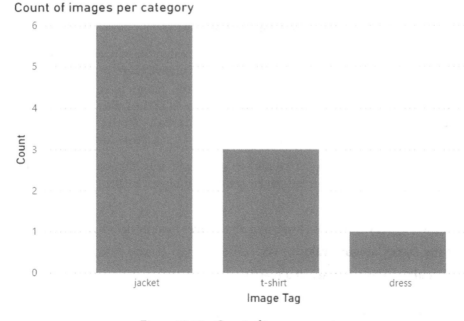

Figure 10.13 – Count of images per category

If we filtered the page by selecting the `dress` column in the stacked column chart, the matrix would filter accordingly, as shown in the following screenshot:

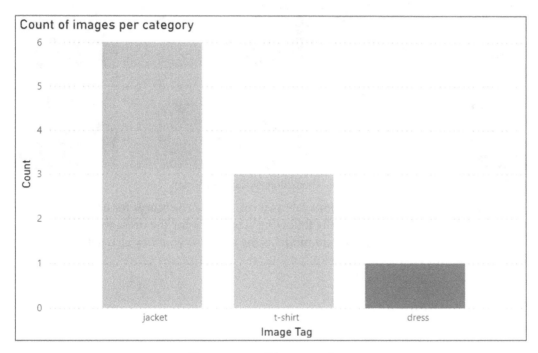

Figure 10.14 – Filtering on dress

From the column chart, we read that only one dress has been found in the images, and in the matrix, only one image of a dress is shown. You may have observed, however, that there was another image of a dress.

Improving the Custom Vision model

To review and improve the model, you can go back to the Custom Vision portal. By opening the **Predictions** tab, as shown in the following screenshot, you can see all images that have been scored by the model through the endpoint:

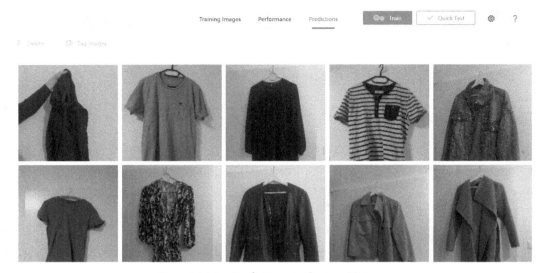

Figure 10.15 – Predictions in Custom Vision

By selecting an image in the **Predictions** tab, you can review the scores for each category on that image, as shown in the following screenshot for the dress we missed. As mentioned earlier in this chapter, no model is the same, so you may also get slightly different probabilities here:

Image Detail   ×

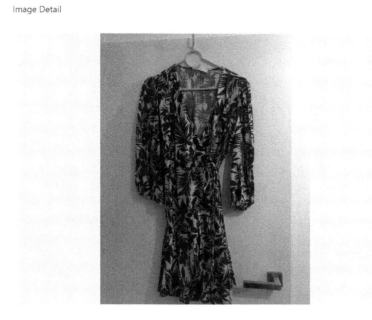

Figure 10.16 – The missed dress

As you can see in *Figure 10.16*, the dress was incorrectly identified as a jacket. A very good cause for this could be our training data. Most images with dresses were of a person wearing them. The model could therefore find it hard to categorize an image as a dress if there were no human in it. To improve the model, there is a quick fix, as outlined here:

1. In the **Image Detail** pane, as shown in *Figure 10.16*, you can add the correct tag under **My Tags**.

2. Select **Save and close**.

3. Do this for all images in the **Predictions** pane.

4. Select **Train** to retrain your model.

5. Review the evaluation metrics of your model. If they are acceptable, publish the newest iteration and unpublish the older iteration to make it unavailable for scoring.

Retraining your model can improve it, and adding more data can improve the model's accuracy. Next to data quantity, think about data quality and diversity when adding more images to your training dataset. Not all images will improve your model. When images show up in the **Predictions** pane, you can choose to either delete them if you expect them not to add any value or use them to help improve your model.

Summary

In this chapter, we have worked with images in several ways. We learned that Power BI understands either the binary content of images or the URLs referring to images. Using the content or URL, we can then use AI Insights to tag images, or we can use the Computer Vision API to tag images. If we want to train a model to use our own labels, we are now able to configure Custom Vision and integrate that with Power BI. And finally, we visualized the images in Power BI to show them next to the insights we had extracted from them. In the next chapter, we'll take a look at how we can customize a model even more by using **Azure Machine Learning (Azure ML)**.

Part 3:
Create Your
Own Models

This section will show you how to have more control over your models by creating them through automated machine learning, Azure ML designer, the Azure ML SDK with Jupyter notebooks, or Visual Studio Code.

This section includes the following chapters:

- *Chapter 11, Using Automated Machine Learning with Azure and Power BI*
- *Chapter 12, Training a Model with Azure Machine Learning*
- *Chapter 13, Responsible AI*

11

Using Automated Machine Learning with Azure and Power BI

Training your own model can be a daunting task. You can try out many different configurations to train a model and identify the best one with **automated machine learning (AutoML)**. **Azure Machine Learning (Azure ML)** allows you to use AutoML for some common ML tasks. After training a model, you can integrate it with Power BI.

Other tools offer similar features to Azure ML; however, we'll focus on Azure ML in this book as the integration of trained models with Power BI is easiest with Azure thanks in part to built-in Power BI features to consume the trained models.

In this chapter, we'll cover the following main topics to get an introduction to AutoML:

- Understanding AutoML
- Creating an AutoML experiment in Azure ML
- Deploying a model to an endpoint
- Integrating the model with Power BI

After going through this chapter, you'll understand what AutoML can be used for. There will still be many features left to discover—this chapter really serves as a first introduction to the concept of using AutoML. Let's go over the technical requirements first.

Technical requirements

To walk through the examples provided in this chapter, you will need the following:

- **Power BI Desktop**

   You can install Power BI from the Microsoft Store or find more advanced downloading options here: `https://www.microsoft.com/en-us/download/details.aspx?id=58494`.

- **An Azure subscription**

   Azure ML is a service available in Azure. To create a service or resource, you need to have access to an Azure subscription. If you don't have one yet, you can sign up for a free subscription here: `https://azure.microsoft.com/en-us/free/`.

- **Sample dataset**

   A sample dataset of the number of tourists per month in the Netherlands is used in the examples provided in this chapter.

   The tourism data is collected by *Statistics Netherlands* and can be found in their online open database here: `https://opendata.cbs.nl/#/CBS/nl/dataset/82058NED/table?searchKeywords=logiesaccommodaties`.

   The dataset also includes weather data, averaged per month, and retrieved from the **Royal Netherlands Meteorological Institute (KNMI)** at `https://www.knmi.nl/nederland-nu/klimatologie/daggegevens`.

   The cleaned-up data that is used in the examples throughout this chapter can be found on GitHub at `https://github.com/PacktPublishing/Artificial-Intelligence-with-Power-BI/blob/main/Chapter11/tourism-extra.csv`.

The dataset to import in Power BI includes data for 2019 on tourism and will be used to test the model in Power BI. You can find it here: `https://github.com/PacktPublishing/Artificial-Intelligence-with-Power-BI/blob/main/Chapter11/tourism-test-2019.csv`.

Understanding AutoML

Azure ML and AutoML may both be new concepts to you. If you do most of your work in Power BI, you may only use these tools occasionally. Multiple books can be dedicated to either of these concepts, which is why we'll cover the bare necessities for data analysts here.

So, why do we want to learn about AutoML? Throughout this book, we have explored many features and services that offer *pretrained models* that are ready to use. There is no need to train them, nor to have the data-science expertise to create models from scratch.

Pretrained models are ideal for common scenarios that many organizations face; for example, one model trained to recognize faces can be used for many different applications. However, if you want to have a forecasting model to predict the demand of your products based on your advertisement strategies to plan the supply, a generic model may not be the right fit for you.

It's when you need the model to be trained and tuned to your data—and to your use case—that you may consider training a model yourself. Even when the requirements call for a new model to be trained, you may not have unlimited resources to create one.

To *minimize the time and effort* needed when training a new custom model, we can use *AutoML*. AutoML is a feature within Azure ML that allows you to train multiple models quickly, after which you can evaluate which best fits your needs. All you have to do is provide data as input and specify what kind of model you need.

Understanding the ML process

Before we dive further into AutoML and how it works, let's review what is needed to train an ML model to understand the process of AutoML. A simplified view of the process to train a model is shown here:

Figure 11.1 – Process to train a model

Although we'll learn in this chapter that AutoML focuses on **feature engineering** (**FE**) and model training to come up with the *best* model, it's good to go over the complete ML process in light of improving the performance of a model.

Improving the performance of an ML model

Ensuring your model performs well can depend on many factors. Within each phase, there are things you can do that may influence the performance of your model, such as these:

- **Define use case**: When deciding on what you want the model to predict, success criteria need to be defined. When is the model good enough to be used in production? The answer to this question will influence the data you'll use, the subset of algorithms you can choose from, and the evaluation metric you'll find most important when drawing conclusions based on the model's performance.

- **Get data**: The dataset needs to be of good quantity and quality. How much data you need depends on things such as the use case, the number of features, and the algorithm you use when training the model.

- **FE**: The dataset can be preprocessed to create features that are expected to influence the prediction. During FE, we can—for example—normalize data to make sure each feature uses the same range of values in order to treat them equally when training the model.

- **Train model**: After deciding which task you want to perform, there are multiple algorithms to choose from. With each algorithm, there are also hyperparameters that can be tweaked to influence the model. For example, you can define a so-called regularization rate value with which you can decide how much you want to punish your model for overfitting on the training dataset.

- **Integrate model**: Finally, after evaluating the model and deciding you want to deploy and integrate it with Power BI (for example), you want to monitor the deployed model to ensure it still performs as needed. If the model underperforms on new data, it may be time to retrain the model on newer data.

When using AutoML in Azure ML, it is still up to you to define a use case and get the data. The purpose of AutoML is to save time and effort when going through FE and model training. By trying different approaches during each phase, AutoML learns which choices will potentially improve the model, optimizing the model quickly without having to do extensive experimentation.

When to use AutoML

With AutoML, some parts of the process of training a model are automated to save time and effort. After providing a training dataset, AutoML can perform *FE* and train multiple models by *choosing different algorithms* each time a new model is trained. For each model, AutoML will *calculate several evaluation metrics* so that you can decide which model best suits your requirements.

You can use AutoML for common ML tasks such as these:

- **Regression**: Predict a numerical value, such as how much a house is worth based on its features.

- **Classification**: Predict a categorical value, such as whether a customer will churn or not.

- **Time-series forecasting**: Predict future values, based on historical data collected at a consistent time interval—for example, predict the energy demand of households per month.

- **Computer vision** (**CV**): Classify images or detect objects in images—for example, detect the variety of apples to automatically sort them.

To train models using AutoML in Azure ML, there are two approaches, as outlined here:

- Using the online **user interface** (**UI**) called **Azure ML Studio**—a user-friendly approach that you can access at `http://ml.azure.com`

- Using the Python **software development kit** (**SDK**)—mostly preferred by data scientists who are already familiar with Python and want to work with the Azure ML service programmatically

As the UI is easier to interpret and understand, we'll use Azure ML Studio for the examples throughout this chapter. Note that everything that can be done through Azure ML Studio can also be done with the Python SDK, with the Python SDK offering even more flexibility and options than the UI.

Now that we know we can use AutoML to automate parts of the process of training ML models, let's see how it works by creating a model with the AutoML feature in Azure ML.

Creating an AutoML experiment in Azure ML

AutoML can be used for training regression, classification, forecasting, or CV models. For the first three, the input dataset is expected to be tabular, but for CV, you'd work with images. To take one task as an example when exploring how to work with AutoML, let's look at forecasting.

In *Chapter 4, Forecasting Time-Series Data*, we already covered forecasting in Power BI. We learned that Power BI can create a forecast for time-series data based on trends and seasonality that it can find in the target value itself, the target value being the number of tourists in the Netherlands per month from 2012 until 2019. Building on that example, we'll add other features to the dataset to train a forecasting model with AutoML that also takes other information into consideration when forecasting the number of tourists.

The dataset we'll use for AutoML has four columns, as follows:

- **StartMonth**: The first day of the month

- **Total tourists**: The total number of tourists in the Netherlands per month—the target value

- **Rain**: Averaged precipitation in **millimeters (mm)**

- **Temperature**: Average monthly temperature measured in °C

The dataset can be downloaded from `https://github.com/PacktPublishing/Artificial-Intelligence-with-Power-BI/blob/main/Chapter11/tourism-extra.csv`, and the first five rows are shown in the following screenshot:

StartMonth	Total tourists	Rain	Temperature
01-01-12	1764000	27.16129	4.880645
01-02-12	1791000	6.758621	0.8
01-03-12	2197000	6.354839	8.293548
01-04-12	3148000	15.73333	8.383333
01-05-12	3629000	27.09677	14.45161

Figure 11.2 – Tourism data

To use Azure ML, we need to create a workspace in Azure. To run AutoML, we'll also need to compute and register the dataset. After that, we can set up an AutoML experiment to train multiple models.

Creating an Azure ML workspace and resources

As the main purpose of this chapter is to see AutoML in action and integrated with Power BI, we'll quickly go through the Azure ML workspace and resources needed to run AutoML.

To train a model in Azure ML, we need the following three artifacts:

- **Azure ML workspace**: Created as a resource in Azure. Represents a centralized workspace in which everything is stored, managed, and monitored.

- **Compute cluster**: A scalable compute managed by the Azure ML workspace. Both **central processing unit (CPU)** and **graphics processing unit (GPU)** types are available in different sizes to accommodate your needs. When idle, it will scale down to zero nodes to save money. When multiple models need to be trained, it can scale down to multiple nodes to train models in parallel.

- **Dataset**: A reference to data stored in cloud storage. Data is ultimately stored in cloud storage such as Azure Blob storage or Azure Data Lake. Datasets contain access information to the data to easily work with data throughout the ML process.

To create an Azure ML workspace, you need access to an Azure subscription. Once you have created a workspace, we'll use Azure ML Studio to create a compute cluster that we'll use to train the models. Finally, we'll register the tourism dataset to use as input for AutoML. Proceed as follows:

1. Open a browser and navigate to `http://portal.azure.com`.

2. Sign in with your account (the one with access to an Azure subscription).

3. In the menu on the left, select **+ Create a resource**.

4. In the search bar that appears, search for `Machine Learning`.

5. Select **Machine Learning** from the search results, as shown in the following
    screenshot:

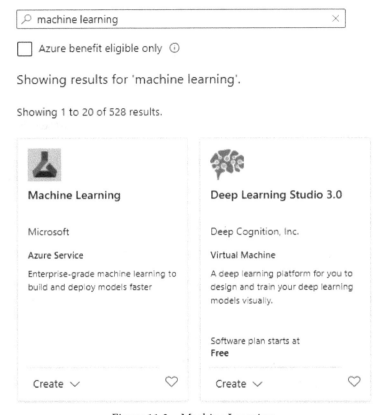

Figure 11.3 – Machine Learning

6. After selecting the card, an overview of the **Machine Learning** service appears.
    Select **Create**.

7. Select the **Subscription** option you want to use (with which you'll pay for the
    resource).

8. Choose an existing **Resource group** type, or create a new one by selecting **Create new**.

9. Enter a value for **Workspace name**.

10. Choose an available region closest to you.

11. Keep the default (to create new) for **Storage account**, **Key vault**, and **Application insights**.

12. Select **Review + create**, and then click again on **Create**.

13. Once all resources have been deployed, go to the **Machine Learning** resource.

14. From the **Overview** tab, select the **Launch studio** button, as shown in the following screenshot:

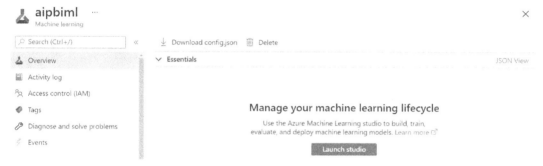

Figure 11.4 – Launching Azure ML Studio

A new tab will open in your browser, to navigate you to `https://ml.azure.com`.

For future reference, you can also access Azure ML Studio by navigating directly to that **Uniform Resource Locator** (**URL**).

15. From the left menu, navigate to **Compute**.

16. Select the **Compute clusters** tab.

17. Select **+ New**, as shown in the following screenshot:

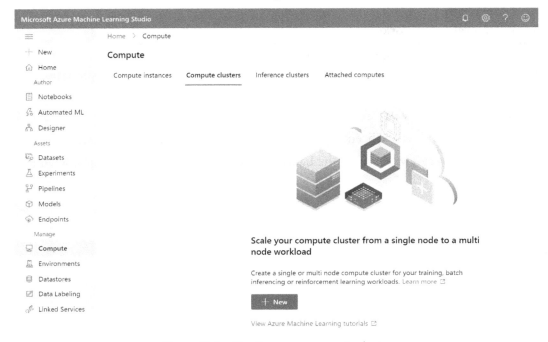

Figure 11.5 – Creating a new compute cluster

18. In the pane that opens, create a new compute cluster with the following settings:

- **Location**: Choose a region nearest to you

- **Virtual machine tier**: Dedicated

- **Virtual machine type**: CPU

- **Virtual machine size**: DS3_v2 (recommended for standard ML workloads; other **virtual machine (VM)** sizes would work too, though they may influence training time)

- **Compute name**: aml-cluster

- **Minimum number of nodes**: 0

- **Maximum number of nodes**: 2

- **Idle seconds before scale down**: 120

A compute cluster will be created for you. When we're not using it for training models, the cluster will be set at zero nodes, meaning we don't pay for compute power. However, we may pay for some storage costs for the metadata.

Next, we need to create a dataset asset in Azure ML to use as input for the AutoML run.

19. From the left menu, select **Datasets**.

20. Select **Create dataset**, as shown in the following screenshot:

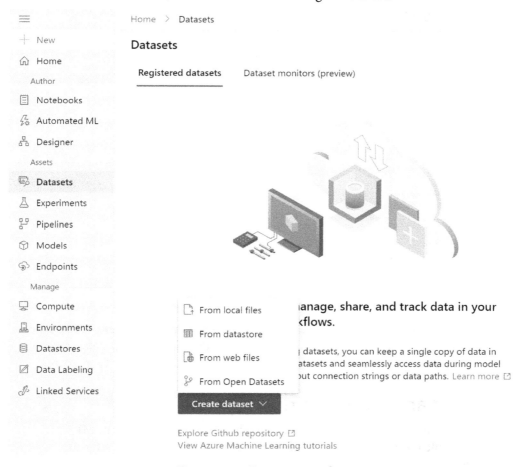

Figure 11.6 – Creating a new dataset

21. From the options provided in the dropdown shown in *Figure 11.6*, select **From local files**.

22. Create a dataset referring to the training data with the following settings:

- **Name**: `tourism-train`

- **Dataset type:** `Tabular`

- Browse to the `tourism-extra.csv` file that you stored locally.

- Keep all settings at their defaults.

- In the **Schema** tab, set the property of `StartMonth` to `Timestamp`, as shown in the following screenshot:

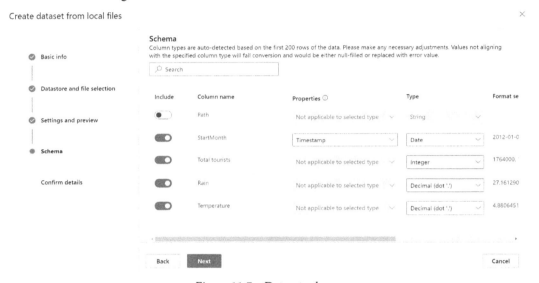

Figure 11.7 – Dataset schema

Now that we have an Azure ML workspace, a compute cluster, and a dataset, we can set up an AutoML run. We'll use the training dataset to train the model. A test dataset is optional. We will use it to see how the model can predict new data and verify its scores.

Configuring an AutoML run

Once all necessary artifacts are in place, we can go ahead and submit an AutoML run to train multiple models. There will be four steps to set up a new AutoML run, as outlined here:

- **Select dataset**: Choose a dataset that meets the requirements, depending on the task you want to use AutoML for. For example, for forecasting, a dataset needs to have a `date` column and the target column needs to be *numeric* (integer or decimal).

- **Configure run**: Each run in Azure ML is part of an *experiment* that allows you to group workloads. Each run will be stored indefinitely in Azure ML and will show all **inputs and outputs (I/Os)** of a workload, such as training a model.

- **Select task and settings**: After choosing the task (classification, regression, or time-series forecasting), you can also decide on settings such as how long you want AutoML to train models, how many models are trained in parallel, and which evaluation metric you want AutoML to focus on to find the *best* model.

- **Validate and test**: To compare models, AutoML will calculate evaluation metrics using the validation method you choose. Optionally, you can specify the test dataset (not used during training) to evaluate the model.

We will set up an AutoML run to create a forecasting model for the number of tourists in the Netherlands. We have data on how many tourists (both Dutch and foreign) were on vacation in the Netherlands per month for the years 2012-2019.

We will take the years 2012-2018 to train the model and use the year 2019 to test the model. To include other features that potentially influence tourism, we also include weather data (average temperature and rainfall).

Let's set up an AutoML run to forecast tourism, as follows:

1. From the left menu, select **Automated ML**.

2. Select **+ New Automated ML run**, as shown in the following screenshot:

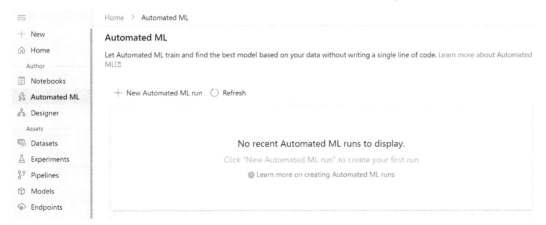

Figure 11.8 – + New Automated ML run

3. When selecting a dataset, choose the `tourism-train` dataset created in the previous section.

4. Configure a run with the following settings:

   - **Experiment name**: `Create new`

   - **New experiment name**: `automl-tourism-forecast`

   - **Target column**: `Total tourists`

   - **Select compute type**: `Compute cluster`

   - **Select Azure ML compute cluster**: `aml-cluster`

5. In the **Task and settings** tab, select **Time series forecasting** with the following configuration:

   - **Time column**: `StartMonth (Date)`.

   - **Time series identifier(s)**: Leave at **Autodetect**.

   - **Frequency**: Turn off **Autodetect** and set the frequency to **Month**.

   - **Forecast horizon**: Turn off **Autodetect** and set the horizon to `12`.

6. Select **View additional configuration settings**. A new pane will open with **Additional configurations** options.

7. Expand the **Exit criterion** section.

8. Set **Training job time (hours)** to `0.5`.

9. Expand the **Concurrency** section.

10. Set **Max concurrent iterations** to 2.

11. Continue to the [**Optional**] **Validate and test** tab.

12. Leave the **validation type** and **number of cross validations** values as they are.

13. For **Test dataset**, leave at **No test dataset required**.

14. Select **Finish** to initiate the AutoML run.

Since we set the concurrency at 2, AutoML will force the compute cluster to scale to two nodes to train two models in parallel. The total time of the AutoML run will likely be a bit more than 0.5 hours (or 30 minutes) because it needs extra time to scale the compute cluster and decide on what the best model was.

Once the AutoML run is completed, you can review the entire run and all the child runs representing the individual models that have been trained.

The **Models** tab will list all models that have been trained and the scores on the primary metric specified, as illustrated in the following screenshot:

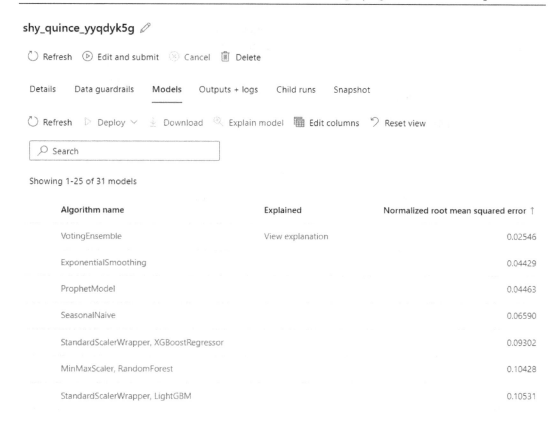

Figure 11.9 – Models trained

You can review each model to also review other evaluation metrics and decide for yourself which model best fits your needs. To use the model in Power BI and get predictions based on data in Power BI, you will need to deploy the best model to an endpoint.

Deploying a model to an endpoint

To get predictions in Power BI, Power BI needs to send data to the model and get the result back to store it as a new column in the dataset. To accomplish this, the model needs to be deployed to a **web service**. When you train a model with AutoML, a web service is very easily created. You only need to specify the following:

- The name of the deployment

- The compute used to generate predictions: either **Azure Container Instances** (**ACI**) for small-scale deployments or **Azure Kubernetes Service** (**AKS**) for large-scale deployments

Both ACI and AKS are container orchestration services. ACI is Azure's proprietary service and is easier to use. AKS is based on the open source Kubernetes technology to orchestrate containerized applications. Even though Azure ML can manage and maintain the AKS clusters used for model deployment for you, it is better to use it when you have the expertise to set it up yourself as the management can be quite complex.

To deploy an AutoML model, follow along with these steps:

1. Select the *best* model of your AutoML run.
2. Select **Deploy**, and then choose **Deploy to web service**.
3. A new pane opens, as shown in the following screenshot:

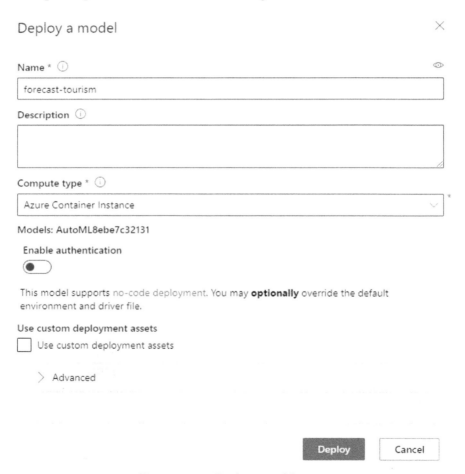

Figure 11.10 – Deploy a model pane

4. Enter a name.

5. Choose **Azure Container Instance** for **Compute type**.

6. Leave **Enable authentication** disabled.

7. Select **Deploy**.

Deployment may take a few minutes. Once it is ready, you'll see a notification at the top. Any live deployments can be found under the **Endpoints** tab.

Now that the model is deployed to a web service, you can call it by using the endpoint. The endpoint works similarly to the endpoints provided by Azure Cognitive Services. To get predictions on new data, you need to make an **application programming interface (API)** call by sending a **HyperText Transfer Protocol (HTTP)** request with data in the body to the endpoint. Power BI has the Azure ML feature Power Query Editor to save your work and make integrating the model easier.

Integrating the model with Power BI

So far, we have been working with Azure ML and AutoML to train a model quickly and easily. Now, it's time to integrate it with Power BI. Proceed as follows:

1. Open a Power BI report.

2. Add the `tourism-test-2019.csv` file to the report. You can download the data from GitHub at `https://github.com/PacktPublishing/Artificial-Intelligence-with-Power-BI/blob/main/Chapter11/tourism-test-2019.csv`.

3. Open the Power Query Editor.

4. Change the data type of the `StartMonth` column to **Date/Time**.

5. From **AI Insights**, select **Azure Machine Learning**.

6. A popup will appear, as shown in the following screenshot. Select the endpoint you created in the previous section:

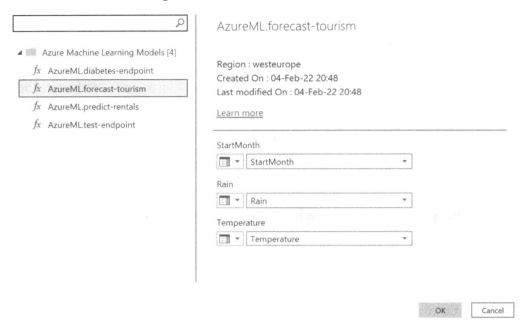

Figure 11.11 – Choosing a model

7. Make sure that each of the columns corresponds to the right input field. The `Total tourists` column is not used to create predictions but is left in the table for us to compare to the predictions.

8. A notification may appear that information is required about data privacy, as shown in the following screenshot. Select **Continue**:

Figure 11.12 – Data privacy

9. Check the box to ignore privacy-level checks, and click **Save**.

Once the data is processed by sending it to the model's endpoint, a new column appears, as shown in the following screenshot:

	StartMonth	1²₃ Total tourists	1.2 Temperature	1.2 Rain	ABC 123 AzureML.forecast-tourism
1	01-Jan-19 00:00:00	2458000	3.506451613	0.350645161	2620087.971
2	01-Feb-19 00:00:00	2565000	6.092857143	0.609285714	2774965.028
3	01-Mar-19 00:00:00	3289000	8.038709677	0.803870968	3139540.025
4	01-Apr-19 00:00:00	4307000	10.92	1.092	3955281.658
5	01-May-19 00:00:00	4193000	11.71935484	1.171935484	4597552.803
6	01-Jun-19 00:00:00	4894000	18.13	1.813	4411331.696
7	01-Jul-19 00:00:00	4704000	18.79032258	1.879032258	4624013.119
8	01-Aug-19 00:00:00	5143000	18.42903226	1.842903226	5057209.125
9	01-Sep-19 00:00:00	4165000	14.53	1.453	4098504.991
10	01-Oct-19 00:00:00	4087000	11.55806452	1.155806452	3998830.795
11	01-Nov-19 00:00:00	3194000	6.366666667	0.636666667	3090453.152
12	01-Dec-19 00:00:00	2918000	5.841935484	0.584193548	2901897.528

Figure 11.13 – Forecasted tourism

Even though we only used two features here, you see in *Figure 11.13* that the predictions are actually pretty close to the actual measurements.

We can plot the actual total tourists and the forecast in the same line chart to get the following output:

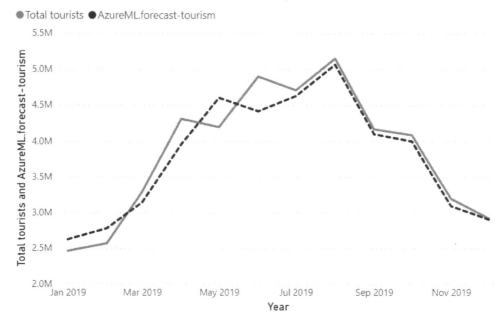

Figure 11.14 – Total tourists compared to forecast

It is to be expected that the forecast is not exactly like the actual data, especially when working with so few features (namely, temperature and rain). Again, many factors influence the model's performance, and we would probably have to improve the training dataset in this case. However, AutoML has managed to give us a good model already by iterating over FE and several algorithms.

Whether you want to train a classification, regression, or time-series forecasting model, all of it can be done with AutoML. All you have to do is make sure your data meets the requirements of the task. Once your model is trained, you deploy it to a web service, after which you can integrate it with Power BI.

Summary

In this chapter, we have explored how to train a model with AutoML, a feature of Azure ML. We created the necessary artifacts: the workspace, the compute, and the dataset. After that, we submitted an AutoML run through the UI and integrated the model with Power BI. In the next chapter, we'll go through the process of training and integrating a model again, but this time, we'll use another feature within Azure ML to do it: the **Designer**.

12

Training a Model with Azure Machine Learning

Throughout this book, we have explored many of the features in Power BI that allow us to apply machine learning without having to train a model. However, you may still encounter a situation where it is necessary to train your model on your data.

To train a model and integrate it easily with Power BI, you can use **Azure Machine Learning** (**Azure ML**). In this chapter, we'll learn how we can use the Azure ML service to create a machine learning model and deploy it to an endpoint. This endpoint can then be integrated with Power BI to get predictions on the data residing in Power BI.

In this chapter, we're going to cover the following topics:

- Understanding how to train a model
- Working with Azure Machine Learning
- Training a model with Azure Machine Learning Designer
- Deploying a model for batch or real-time predictions
- Integrating an endpoint with Power BI to generate predictions

As Azure ML may be a new topic to you, we'll also walk through an example of training a model with the Designer and integrating that with Power BI. To follow along with this walkthrough, please review the *Technical requirements* section to ensure you have the necessary prerequisites.

Technical requirements

There are three things you will need to go through the examples provided in this chapter:

- Power BI Desktop:

  As this book revolves around Power BI, we expect you to have Power BI Desktop installed on your system. You can install Power BI from the Microsoft Store or find more advanced download options here: `https://www.microsoft.com/en-us/download/details.aspx?id=58494`.

- An Azure subscription:

  Azure ML is a service that's available in Azure. To create a service or resource, you need to have access to an Azure subscription. If you don't have one yet, you can sign up for a free subscription here: `https://azure.microsoft.com/en-us/free/`.

- Sample dataset for Power BI:

  A sample dataset on self-rated happiness for countries from 2005 to 2020 will be used in this chapter's examples. More information on this dataset can be found in *Chapter 2, Exploring Data in Power BI*. You can download the full dataset here: `https://github.com/PacktPublishing/Artificial-Intelligence-with-Power-BI/tree/main/Chapter12/world-happiness-report.csv`.

- Training and test dataset for Azure ML:

  The world happiness dataset contains 36 rows where `Log GDP per capita` is missing. To train a model that can impute these missing rows, we have created a training dataset that excludes missing rows. You can find the training dataset here: `https://github.com/PacktPublishing/Artificial-Intelligence-with-Power-BI/tree/main/Chapter12/happiness-train.csv`. Another dataset includes two of the 36 empty rows. You can find that dataset here: `https://github.com/PacktPublishing/Artificial-Intelligence-with-Power-BI/tree/main/Chapter12/happiness-missing-gdp.csv`.

Understanding how to train a model

Azure ML is designed to help (citizen) data scientists train, manage, and monitor machine learning models. The Azure ML workspace is a platform that contains a variety of features to help you during the machine learning process. Understanding the need for each feature in Azure ML will help you understand the machine learning process.

Understanding the machine learning process

We covered the machine learning process previously in this book in *Chapter 1, Introducing AI in Power BI*, and *Chapter 11, Using Automated Machine Learning with Azure and Power BI*. The overview that was provided in those chapters is shown in the following diagram:

Figure 12.1 – Process to train a model

As we will be going through each step while training a model with Azure ML, let's review each phase.

Defining the use case

At the start of any machine learning project, there should be a discussion with all the stakeholders. What is the purpose of the model you want to train? What would you consider to be a successful model? And, of course, what should the model predict? Many different techniques can be used to train a model.

The three most common use cases are as follows:

- **Regression**: Predict a *numerical* value – for example, what the price of a house is likely to be, based on the house's features.

- **Classification**: Predict a *categorical* value – for example, whether a customer will churn or not.

- **Clustering**: Divide your data into *groups*. Within each group, the features should be similar. For example, you can divide your customers into clusters. Each cluster can represent a type of customer or a customer's persona.

Deciding on the purpose of the model will help you identify what data you need to find or collect to train the model.

Get data

To train a model with Azure ML, you need to have data. If you're lucky, you will already have the data you need to train the model you want. In other cases, you may need to collect the data or buy the data from another organization.

Data can be stored in a variety of databases, depending on its format. When you're training a model, the data needs to be stored as delimited (for example, in a CSV file), Parquet, JSON, or plain text. For Azure ML to work with the data, it is best to store the data in any of these formats in an Azure cloud solution such as Azure Blob Storage or Azure Data Lake.

Once you have connected the data to Azure ML, it is time to explore it, as we did in *Chapter 2, Exploring Data in Power BI*. At this stage, you'll learn whether you have any missing data or missing features to help create the model.

Feature engineering

Data has a major influence on the performance of the model. The quantity and quality of the data will matter when you're training a machine learning model. To improve its quality, you can exclude columns or rows. The columns or fields that will act as the input for the model are referred to as features in data science. This is also why this phase is often referred to as **feature engineering**: we're manipulating the input of the model.

During feature engineering, you'll want to **remove any high-cardinality features** as they rarely provide any useful information to the model. High-cardinality features are fields that include many unique values. Almost every row or all rows will contain a unique value. For example, the customer ID is considered a high-cardinality feature. Features such as customer ID should be removed.

Next to that, you'll also want to **clean missing data** by, for example, removing the complete rows with missing values or imputing missing data with the mean or median of that column.

The values of a feature can also influence a model. Let's look at an oversimplified example. Let's say that we want to predict customer churn and that we have two features, among others: how many of your luxury products customers have bought in the past and the price of those products. Your customers may tend to buy few products as prices are high.

However, even though the values of the `price` feature are much higher than that of the `number of products sold` feature, you want both features to be equally important to the model. To equalize them, you can **normalize** the data. During normalization, each field is put on a similar scale, maintaining the distribution of the data to ensure the relative values matter, not the absolute values.

Finally, a data scientist may manipulate the data to create new features out of old ones. This can be done to simplify the dataset, as well as to remove unnecessary details that may distract the model. For example, when you're predicting customer churn, you may have a dataset that includes the home address of each customer. The home address itself is too detailed and will not be useful for the model. Instead, it is potentially more interesting to know how close a customer lives to your stores. In that case, you can use the home address to calculate the `distance to store` feature and create a new feature for that.

Similar to the complete machine learning process, feature engineering is an iterative endeavor. Testing different ways to present the data as input when you're training a model will teach you what the best dataset is. Some common data manipulations that occur during feature engineering include removing high-cardinality features, cleaning missing data, and normalizing features.

Once you feel like you have a good enough dataset, you're ready to train the model.

Training the model

Once you are reading to train the model, you will have a lot of information. You should know what model you want to train, and you should have gone through feature engineering to create a quality dataset to present as input. Besides a dataset, you'll need to decide on an algorithm to train the machine learning model, based on the task you want to perform.

Which algorithm is available to you depends on what approach you take when working with Azure ML. If you are already an experienced data scientist, you can likely continue using the open source frameworks you were using before working with Azure ML. If you aren't an experienced data scientist, then don't worry – Azure ML offers a few common algorithms for common use cases in the workspace. We'll explore these options in the *Training a model with Azure Machine Learning Designer* section.

Once you've trained the model, you'll need to evaluate it to review its performance. Very often, a validation or test dataset is used to evaluate the model. The benefit of using a validation or test dataset is that you know what the target value should be. So, when you're using the model to create predictions on the test dataset, you can compare the actual values with the predicted values to evaluate the model's performance. How you evaluate these depends on the task and the algorithm that was selected during training. What you consider a good enough performance depends on the success criteria for the model.

Integrating the model

One challenge with machine learning has been bringing models to production. Bringing a model to production means making the trained models available where they need to be used to create predictions. Azure ML facilitates this by creating an **endpoint** for the model. This endpoint allows you to integrate the model with any application you want.

To integrate the model, you'll need to deploy it. To deploy the model, you need to decide whether you want one of the following:

- **Batch predictions**: Generate new predictions asynchronously on a new dataset that contains multiple data measurements. The predictions will be generated based on a schedule or a trigger and stored in a data store.

- **Real-time predictions**: Generate a new prediction for each new data measurement. As the predictions are calculated for each measurement, they can be generated in real time: when you send a new data point to the endpoint, you'll get a result almost instantly.

In general, batch predictions can be more cost-efficient. You can integrate the generations of batch predictions in a data transformation pipeline upstream. If, for example, you have a data pipeline to transform data before it enters Power BI, you can integrate Azure ML with that pipeline to generate batch predictions and add them to the data before it gets loaded into Power BI.

Adding batch predictions upstream is the optimal decision when you're working with large amounts of data. An example of a service that can integrate with Azure ML to create such a pipeline is Azure Data Factory (an ETL tool), which is also part of Azure Synapse Analytics (a data warehousing solution).

To integrate a model trained in Azure ML directly with Power BI, you must use real-time predictions. By deploying a model to a real-time endpoint and integrating that with Power BI, you can send data from Power BI to the model and get a prediction back within Power BI as well.

Although real-time endpoints are great for these kinds of immediate predictions, it is more compute-heavy for Power BI. If you have a large dataset in Power BI that you want to generate predictions for, it may take some time before each row is sent to the endpoint and the result is stored in a new column. If that time is negatively influencing your Power BI report's performance, think about generating the predictions upstream.

With that, we have revisited the phases of the machine learning process and know what to expect when training a model in Azure ML. In the next section, we'll cover some of the fundamental concepts of Azure ML and, most importantly, how to work with the Azure ML service.

Working with Azure ML

Learning how to work with Azure ML is worth multiple books by itself. Azure ML is designed to help data scientists create and manage machine learning models that can be trained with the often-preferred Python framework.

In this section, we'll cover the basics of the Azure ML service that may interest you as a data analyst. We'll also touch upon some concepts within the service that are reserved for more advanced users so that you know what to explore further.

There are three approaches to working with Azure ML:

- Using the online UI called **Azure Machine Learning Studio**. This is a user-friendly approach that you can access at `http://ml.azure.com`.

- Using the *Python SDK*. This is mostly preferred by data scientists who are already familiar with Python and want to work with the Azure ML service programmatically.

- Using the *Azure CLI extension for Azure ML*. This is mostly preferred by administrators and machine learning engineers who want to automate tasks such as creating the workspace and the necessary resources (such as compute).

For feature engineering, training a model, and integrating that model by deploying it, you can use Azure ML. For each of these phases of the machine learning process, we can work with the following Azure ML assets:

- **Datasets**: By creating a dataset, Azure ML learns how to connect to a specific file or files. We'll create a dataset when we upload data to the Azure ML workspace's default datastore to train and test the model.

- **Compute cluster**: A scalable compute cluster that's managed by the Azure ML workspace and designed for data science workloads. When idle, it will scale down to zero nodes to save money.

- **Experiments**: Each time we execute a workload such as model training in Azure ML, it will be tracked as a new run within an experiment. The experiment runs are stored indefinitely and allow you to review all the inputs and outputs of the workloads that are submitted in Azure ML.

- **Endpoints**: This provides an interface for generating predictions on new data by a trained model. Deploying a model that's been trained in Azure ML to an endpoint will allow us to integrate the model with Power BI.

You can create an Azure ML workspace and assets through the UI, Python SDK, or Azure CLI. In this chapter, we'll use the UI as it is easiest to grasp for novices. However, all these steps can be reproduced in a Python SDK and the Azure CLI.

To train a model and deploy it to an endpoint, and then integrate the model with Power BI, follow these steps:

1. Create an Azure ML workspace.

2. Create an Azure ML compute cluster.

3. Create an Azure ML dataset.

4. Create a new pipeline with the Designer in the Studio (the alternative to this when using the Python SDK or Azure CLI would be a training script).

5. Train and evaluate the model (by executing the pipeline or training script).

6. Create an inference pipeline (alternatively, an inferencing or scoring script).

7. Deploy the model that contains the inference pipeline to an endpoint.

8. Integrate the endpoint with Power BI using AI Insights's **Azure ML** feature.

> **Use an organizational account**
>
> To integrate an Azure ML endpoint with Power BI, you need to sign into Power BI with the account that has access to the endpoint. Though you can use a personal account to attach to an Azure subscription, you can only sign into Power BI Desktop with an organizational account. Ensure you use an organizational account when you're deploying the model to an endpoint in Azure ML to make the endpoint available in Power BI.

For clarity, these steps have been divided into several sections. First, we'll start with the three first steps to create the necessary Azure ML assets.

Creating Azure ML assets

To create an Azure ML workspace, you need access to an Azure subscription. Once you have created the workspace, we'll use Azure ML Studio to create a compute cluster that we'll use to train the models. Finally, we'll register two datasets that include the world happiness measures we worked with in *Chapter 2, Exploring Data in Power BI*. We'll discuss the data when we train the model in the next section.

Let's get started:

1. Open a browser and navigate to `http://portal.azure.com`.

2. Sign in with your account (an *organizational* account with access to an Azure subscription).

3. From the menu on the left, select **+ Create a resource**.

4. In the search bar that appears, search for **Machine Learning**.

5. Select **Machine Learning** from the search results.

6. After selecting the card, an overview of the **Machine Learning** service will appear. Select **Create**.

7. Select the **Subscription** option you want to use (which you'll pay for the resource with).

8. Choose an existing **Resource group** or create a new one by selecting **Create new**.

9. Enter a **Workspace name**.

10. Choose an available **Region** that's closest to you.

11. Keep the defaults (to create new) for **Storage account**, **Key vault**, and **Application insights**.

12. Select **Review + create**, and then **Create**.

13. Once all the resources have been deployed, go to the **Machine Learning** resource.

14. From the **Overview** tab, select the **Launch studio** button. A new tab will open in your browser that will navigate you to `https://ml.azure.com`.

15. From the left menu, navigate to **Compute**.

16. Select the **Compute clusters** tab.

17. Select **+ New**.

18. In the pane that opens, create a new compute cluster with the following settings:

    -   **Location**: Choose the region nearest to you.

    -   **Virtual machine tier**: `Dedicated`.

    -   **Virtual machine type**: `CPU`.

    -   **Virtual machine size**: `DS3_v2` (recommended for standard machine learning workloads; other VM sizes would work too, though they may influence training time).

- **Compute name**: `aml-cluster`.
- **Minimum number of nodes**: `0`.
- **Maximum number of nodes**: `2`.
- **Idle seconds before scale down**: `120`.

The compute cluster will be created for you. When we're not using it to train models, the cluster will be set at zero nodes, meaning we don't pay for compute power. However, we may pay for some storage costs for the metadata.

Next, we need to create two dataset assets in Azure ML to use as input for the model training phase and to test the model before deploying it.

19. From the left menu, select **Datasets**.

20. Select **Create dataset**.

21. From the options provided in the dropdown, select **From local files**.

22. Create a dataset that refers to the training data by using the following settings:

- **Name**: `happiness-train`.
- **Dataset type**: `Tabular`
- **Browse** to the `happiness-train.csv` file that you stored locally.
- Keep all the settings as their defaults.
- In the **Schema** tab, review all the columns and their types. Note that `Country name` is the only column that contains string data.

23. Create a second dataset that refers to the test dataset by using the following settings:

- **Name**: `happiness-missing-gdp`.
- **Dataset type:** `Tabular`.
- **Browse** to the `happiness-missing-gdp.csv` file that you stored locally.
- Keep all the settings at their defaults.
- In the **Schema** tab, review all the columns and their types. Note that the test dataset doesn't include the `Log GDP per capita` column.

Now that you have an Azure ML workspace that contains a compute cluster and two datasets, you're ready to train the model in Azure ML through the UI, Python SDK, or the Azure CLI.

Training a model with Azure ML Designer

As a data analyst, you may be an aspiring data scientist, but not yet fluent in creating training scripts written in Python while using common libraries such as scikit-learn to train a model. Therefore, we'll keep things approachable and train a model with the visual drag-and-drop interface that's provided by Azure ML: the **Azure ML Designer**.

Remember that everything that you can do with the Designer can also be replicated by creating scripts and running them with the Python SDK or the Azure CLI.

When you want to train a model with the Designer, there are several common algorithms you can choose from. With these built-in components, you can easily train a regression, classification, or clustering model. In this section, we will train a regression model on the world happiness dataset to fill any empty rows that we have and fix missing data more intelligently than by simply using a mean or median.

In *Chapter 2, Exploring Data in Power BI*, we worked with the world happiness dataset, which contains data on countries' happiness per year. To rank each country based on how happy their inhabitants are, we can review the Life Ladder score, which varies between 1-10, with 10 being the happiest a country's population can be. Additionally, to elaborate on why countries may score low or high on the Life Ladder score, other measures have been collected for each country:

- Log of Gross Domestic Product (GDP) per capita: A measure of how much a country produces, which is often used as an indicator of a country's economic health. This log is used to make it easier to compare the levels between countries.

- Social support: The national average of responses to survey questions related to whether respondents felt like they can get support from their social circles, such as family and friends.

- Healthy life expectancy at birth: This is based on data from the World Health Organization.

- Freedom to make life choices: The national average of responses to survey questions related to whether people felt like they had the freedom to make their own life choices.

- Generosity: The national average of responses on survey questions asking whether participants have donated money to charity recently.

- Perception of corruption: The national average of responses on survey questions related to corruption throughout government and within businesses.

After importing the world happiness dataset into Power BI, you may notice that there is some missing data. Tools such as **Column quality** in **Power Query Editor** can quickly show which columns contains missing data, or empty rows, as shown in the following screenshot:

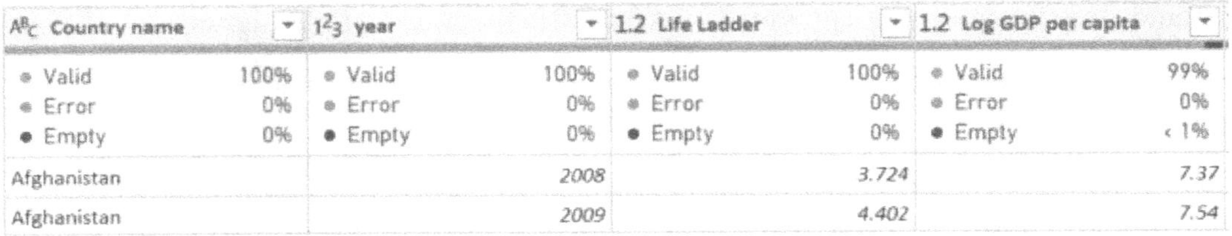

$^{ABC}_C$ Country name		$^{123}_3$ year		1.2 Life Ladder		1.2 Log GDP per capita	
● Valid	100%	● Valid	100%	● Valid	100%	● Valid	99%
● Error	0%	● Error	0%	● Error	0%	● Error	0%
● Empty	0%	● Empty	0%	● Empty	0%	● Empty	< 1%
Afghanistan			2008		3.724		7.37
Afghanistan			2009		4.402		7.54

Figure 12.2 – Empty rows in the Log GDP per capita field

As we discussed in *Chapter 3*, *Data Preparation*, there are several ways to fix missing data. You can use the mean or median with numerical data to fill the empty rows or use the mode or most frequent value for categorical data. Sometimes, you may want to use a more intelligent approach.

To impute missing data, you can also use a machine learning model. In the case of the world happiness data, less than 1% of the rows are empty for Log GDP per capita. Since the GDP per country can differ greatly and influences the Life Ladder score, which is what we're ultimately interested in, we can opt to impute values by using a model instead of using a mean or median.

The rest of this chapter will use this example to show you how to create a machine learning model. Note that you can use machine learning models to impute missing values, as well as to predict unknown values when there is no missing data. However, samples for using machine learning for predicting customer churn, for example, are aplenty. As a data analyst, you're highly likely to have to deal with missing data, which is why we're using this example here.

In the next section, we'll configure an Azure ML Designer pipeline to train the model. In the previous section, you created two datasets in Azure ML:

- The happiness-train dataset: The subset of the world happiness dataset that Log GDP per capita is measured for (11 columns, 1,913 rows).

- The happiness-missing-gdp dataset: The subset of the world happiness dataset that Log GDP per capita is missing for (10 columns, 2 rows).

To train a model, you will preprocess the `happiness-train` dataset. You will do any necessary feature engineering, such as fix missing data in any of the other columns, as well as normalize the `Healthy life expectancy at birth` column since its range is much larger than the other features.

You will split the training dataset so that 75% of the dataset can be used to train the model and the other 25% of the dataset can be used to test the model. Splitting the data is a common practice to test the model and anything around a 75/25 split is often used.

We'll train the model by choosing one of the regression algorithms as we're predicting a numerical value. Feel free to experiment with different algorithms to see how they'll influence the model's performance.

Once the model has been trained, we'll test the model with the 25% of the dataset where we know what the `Log GDP per capita` score is supposed to be, to evaluate how well the model performs.

After evaluating the model, we'll deploy it by creating a scoring or inference pipeline. To test the inference pipeline, we'll use the `happiness-missing-gdp` dataset with the *empty rows* for `Log GDP per capita` to see the model in action before we integrate it with Power BI to add the predicted values to our dataset.

Let's start by creating the Azure ML Designer pipeline, which will load in the training dataset, perform feature engineering, train the model, and evaluate the model.

Configuring an Azure ML Designer pipeline

We'll create a new pipeline with the Designer and execute it with the compute cluster we created in the previous section. The input for the pipeline will be the training dataset, including the world happiness measures.

We'll fix any missing data by replacing empty rows with the median. Alternatively, you could use the mean to fill the empty rows. However, since many of the features have a skewed distribution (see *Chapter 2, Exploring Data in Power BI*), we'll use the median. We'll normalize the `Healthy life expectancy at birth` column. Once we have split the data, we can train a regression model to predict the `Log GDP per capita` score.

Let's start by creating a new pipeline in the Designer:

1. Navigate to Azure ML Studio at `https://ml.azure.com`.

2. From the left menu, select **Designer**, as shown in the following screenshot:

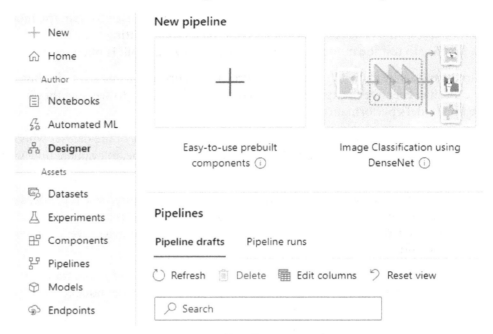

Figure 12.3 – New Designer pipeline

3. Select the + symbol to create a **new pipeline**. A new blank pipeline will be created.

   As shown in the following screenshot, on the left, you'll have the components you can drag and drop onto the canvas to create a pipeline. On the right, a pane will appear when a component is selected so that you can configure its functionality. At the top, the title of the pipeline will be shown, indicated by a cogwheel symbol, so that you can open the pipeline's settings on the right.

4. Change the generated pipeline's name to `Train-model-log-GDP`.

5. Select the cogwheel next to the pipeline's name to open the pipeline's **Settings**.

6. In the **Settings** pane, change the compute type to **compute cluster**.

7. Select the Azure ML compute cluster you created in the previous section; that is, `aml-cluster`. The pipeline should look as follows:

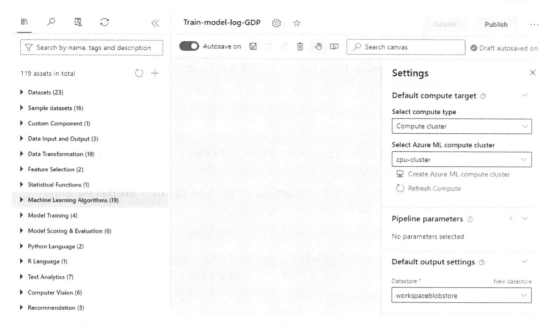

Figure 12.4 – Settings for the new pipeline

8. Expand the **Datasets** category from the assets presented on the left to find the datasets you registered in the previous section.

9. Drag and drop the **happiness-train** component at the top of the canvas.

   You can easily find other components you need to add by searching for their names at the left top.

10. Add the **Clean Missing Data** component and connect its input to the output of the **happiness-train** component. You can connect components by selecting an output, holding down on what you've selected to create a connection line, and dragging it to the next component's input.

11. In the configuration pane for **Clean Missing Data**, edit the columns to be cleaned. Select the following columns:

    - `Healthy life expectancy at birth`

    - `Freedom to make life choices`

    - `Generosity`

    - `Perceptions of corruption`

    - `Positive affect`

    - `Negative affect`

12. For **Cleaning mode**, select **Replace with median**.

13. Add a **Normalize Data** component.

14. Connect the **Clean Missing Data** output to the **Normalize Data** input.

15. Edit the columns to be transformed and select `Healthy life expectancy at birth`.

16. Add a **Split Data** component.

17. Connect the left output port of the **Normalize Data** component to the input of the **Split Data** component.

18. Edit the **Split Data** component. For **fraction of rows in the first output dataset**, change the default value to `0.75`.

19. Set **Random seed** to `123`.

    The left output port of the **Split Data** component will contain 75% of the rows of the happiness dataset. The right output port will contain the other 25% of the rows .

20. Add a **Train Model** component.

    The left input port of the **Train Model** component expects an algorithm. The right input port expects the training dataset.

21. Connect the left output port of **Split Data** to the right input of the **Train Model** component.

22. From the available components on the left, expand the **Machine Learning Algorithms** section; you'll see six options to train a **regression** model.

23. Select the **Decision Forest Regression** algorithm component and add it to the canvas.

24. Add the output of the algorithm to the left input of the **Train Model** component.

25. Edit the **Train Model** component and select the `Log GDP per capita` column for **Label column**.

26. Add a **Score Model** component.

27. Connect the **Train Model** output to the left **Score Model** input.

28. Connect the right output of **Split Data** to the right input of **Score Model**.

29. Add an **Evaluate Model** component.

30. Connect the **Score Model** output to the **Evaluate Model** input.

The final pipeline should look as follows:

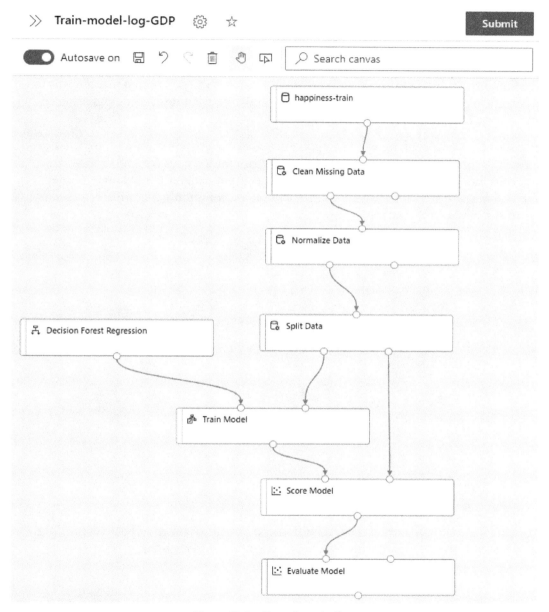

Figure 12.5 – Complete pipeline

31. Select **Submit** at the top right to run the pipeline.

32. A pop-up will appear. For **Experiment**, select **Create new**.

33. Enter train-GDP-model for **New experiment name**.

34. Select **Submit** to run the pipeline.

After submitting the experiment run, the compute cluster will be notified that it needs to execute the pipeline. Since we're not doing any parallel computing, the compute cluster will only use one node to execute the run.

Once the run is executed, for each component, you can see whether it was completed successfully. There is one overarching experiment run for the complete pipeline, and child runs for each component. If you preview the output data for the **Evaluate Model** component, you'll see some evaluation metrics that are commonly used with regression algorithms, as shown in the following screenshot:

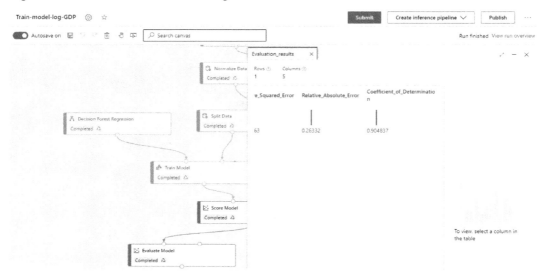

Figure 12.6 – Evaluating the regression model

For example, the **coefficient of determination** component is shown to be **0.904837**. The closer it is to 1 instead of 0, the more accurate the model is likely to be.

Another component that we can add to evaluate the model is the **Execute Python Script** component. By taking the predicted and actual values, as well as the output of the **Score Model** component, we can plot these two values against each other in a *scatter plot* using the following code in an **Execute Python Script** component:

```python
import pandas as pd
import matplotlib.pyplot as plt


def azureml_main(dataframe1 = None, dataframe2 = None):

    # Get actual and predicted values
    actual = dataframe1[['Log GDP per capita']]
    predicted = dataframe1[['Scored Labels']]

    # Plot outputs scatter plot
    plt.scatter(actual, predicted,  color='black')

    plt.xlabel('Actual Log GDP per capita')
    plt.ylabel('Predicted Log GDP per capita')
    plt.title('Scatter plot')

    img_file = "Scatter.png"
    plt.savefig(img_file)

    from azureml.core import Run
    run = Run.get_context(allow_offline=True)
    run.upload_file(f"plots/{img_file}", img_file)

    return dataframe1
```

When the Python script to create and store the scatter plot is connected to the output of the **Score Model** component, you'll find the plot shown in the following screenshot under the **Images** tab of the **Execute Python Script** component:

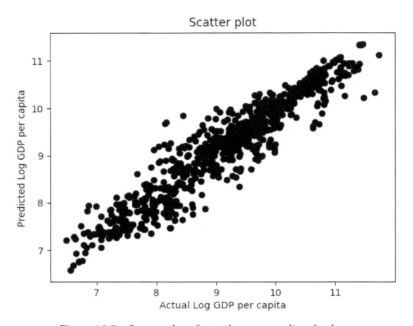

Figure 12.7 – Scatter plot of actual versus predicted values

The more the dots in the scatter plot are aligned in a diagonal line going from the bottom left to the top right of the chart, the more accurate the model is likely to be. In other words, the predicted values are close to the actual values.

> **Evaluation metrics depend on algorithm used**
>
> Which evaluation metrics you use to assess the model's performance depend on the algorithm you select during training and the requirements of the model. The purpose of this chapter is to show you how evaluation metrics can be obtained in various ways. It doesn't provide an extensive list of metrics for each algorithm you can use.

You can use the Azure ML Designer to train a model using commonly used algorithms. Using built-in components in the Designer, you can perform feature engineering to improve the data's quality before training the model. Finally, you can evaluate the model by either using the metrics that are calculated automatically for you, using Python scripts to calculate metrics, or creating plots that help you decide whether the model is ready to use and be deployed.

Remember that everything we did in the Azure ML Designer can also be done by creating Python scripts. The algorithms that are available in the Designer are based on frameworks such as scikit-learn, which is commonly used by data scientists. If you choose to train a model by creating Python scripts, you're more likely to run the script by using the Python SDK or the Azure CLI extension that's been designed for the Azure ML workspace. Whatever approach you use, you have seen that we can train a model with it. In the next section, we will deploy the model.

Deploying a model for batch or real-time predictions

You can train a model anywhere, so why bother using Azure ML? Mostly because it gives us a very easy way to deploy the model to integrate it with Power BI. Before we get into how to deploy the model with Azure ML, let's go over some theory behind why we would want to deploy a model at all.

Training a model can cost a lot of compute power. The reason why AI has had a resurgence is due to the easy availability of cloud compute, among other reasons. Instead of needing a computer with enough processing power, we can now simply use whatever compute is necessary from any cloud provider to train the model, and not use it when we're done training the model.

By having more compute at our disposal, we can train models with complicated algorithms, but also use more data, which generally improves a model's accuracy.

However, training a model and using that trained model to generate new predictions, doesn't require the same type of compute per se. Generating new predictions is called **inferencing** in Azure ML and when it comes to infrastructure, the best approach depends on whether you want to get *batch* or *real-time* predictions.

Generating batch predictions

Although every use case is different, most often, the requirements ask for batch predictions. This means that in the data transformation or cleansing pipeline that is happening upstream from Power BI, you also want to generate predictions with the machine learning model you have created. These predictions are stored alongside the original data and then loaded into Power BI.

Just like any transformations you want to perform on the data, Power BI performs best when data processing is happening as close to the data source as possible. If we're thinking of keeping everything within the **Azure** platform, that could mean our data pipeline looks like this:

1. Transactional data coming from cash registers in shops stored in an **Azure SQL Database**.

2. **Azure Data Factory** or **Azure Synapse Analytics** is used to load the data from the *Azure SQL Database*.

3. The data is cleaned and transformed according to the requirements of the data analyst.

4. The cleaned and transformed data is again stored in another table of the *Azure SQL Database*.

5. *Azure Data Factory* or *Azure Synapse Analytics* initiates an **Azure ML pipeline**.

6. The *Azure ML pipeline* loads the trained model, generates predictions, such as what the sales will be per product in the coming week, and stores the forecast in another table in the *Azure SQL Database*.

7. *Power BI* loads the data from the tables in the *Azure SQL Database*. Once the data has been imported, you can use that data, including the predictions, to build reports.

If you are planning to refresh the data you loaded into Power BI regularly, and you want to include new predictions at a similar cadence, you should use batch predictions. To generate batch predictions with Azure ML, you can create a pipeline that can partition the new dataset or table and run the batch predictions in parallel.

Once you have trained a model with the Azure ML Designer, you can easily create a batch inferencing pipeline by selecting the **Create inference pipeline** button at the top right of the canvas. You'll want a pipeline that looks similar to the following:

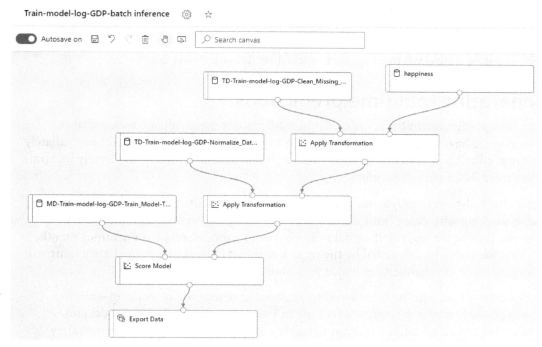

Figure 12.8 – Batch inferencing pipeline

As you can see, new data is loaded at the top of the pipeline (such as from an Azure SQL Database), is transformed to resemble the training data (including fixing missing data and normalizing features), and is scored by using the trained model. Finally, the predicted values are exported and stored again in a database (such as the Azure SQL Database).

For batch predictions, you're likely to create such a pipeline, whether it's created with the Designer, with the Python SDK, or with the Azure CLI. If you want the data to be predicted to be split up and scored in parallel, you can use a compute cluster in Azure ML to parallelize the work and reduce processing time. The benefit of using a compute cluster is that it will scale up and down automatically so that it's cost-efficient.

Whenever you want to get batch predictions, create an inferencing pipeline and execute it with an Azure ML compute cluster. To initiate the pipeline, you can publish the pipeline to schedule it from within Azure ML. Azure ML also generates an endpoint for a published pipeline, which will allow you to trigger the pipeline from any other application, such as Azure Data Factory. Just remember that no matter where you trigger the pipeline from, the predictions will be stored in the database that's specified in the pipeline itself.

For many use cases, a batch inferencing pipeline will best meet your requirements while still maintaining good performance in Power BI. Alternatively, to avoid integrating a batch inferencing pipeline with our upstream data processing pipeline, we can use real-time inferencing in Azure ML to integrate models directly with Power BI.

Generating real-time predictions

Real-time prediction, in the context of Azure ML model deployments, means that whenever we have data, we want to be able to trigger an endpoint to (almost) immediately get the prediction or result back. Real-time deployments allow models to directly integrate with Power BI for ease of usability.

Normally, real-time deployments are used whenever you're using an application where you're working with data. Think about a website selling clothes, for example. Based on how you browse through clothes, data can be sent to a model, recommendations on other clothing items can be calculated by the model, and that data can be immediately returned to you through the website to enhance your shopping experience.

Since Power BI can also be considered an application where you're working with data, we can go for a similar approach here. Data in Power BI can be sent to a model and predictions can be calculated and immediately returned to Power BI, after which they are stored in the dataset that's been created by Power BI.

To deploy a model to a real-time endpoint and meet the expectation of returning predictions in real time, you will want a more lightweight compute than when you're calculating batch predictions. Instead of creating a pipeline that processes a subset (for example, a table with many rows) to generate batch predictions, we now want a pipeline that processes only one or two rows of data. These real-time predictions are often calculated by using **containers** as compute, as they are ideal for such scenarios.

Creating the real-time inference pipeline

When you use the Azure ML Designer to train a model, you can create a real-time inferencing pipeline by following these steps:

1. From the training pipeline that you created in the previous section in the Designer, select the **Create inference pipeline** button at the top right of the canvas.

2. Delete the **Evaluate Model** component (and the **Execute Python Script** component if you added it).

3. Delete the `happiness-train` dataset component and replace it with the `happiness-missing-gdp` dataset.

4. Connect the `happiness-missing-gdp` dataset to the right output of the first `Apply Transformation` component.

When you run the pipeline in the Designer to test whether it works, it will use the `happiness-missing-gdp` dataset to run through all the components. Only two rows of data have been included because we want to test whether the pipeline works. After deployment, it will take the input it gets when using the endpoint, which is referred to with **Web Service Input**. With this endpoint, we'll integrate the model with Power BI to get predictions on all 36 missing rows.

As a result, you should get a pipeline that expects a **Web Service Input** (new data with the same schema as your training data) and a **Web Service Output** (returned predictions), as shown in the following screenshot:

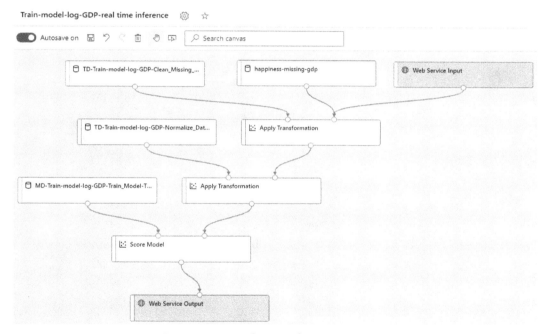

Figure 12.9 – Real-time inferencing pipeline

5. Test the pipeline by selecting **Submit**.

6. Set up the pipeline run by creating a **new experiment** named `test-inference-gdp`.

Once the real-time inference pipeline has finished executing, you can check the predictions by previewing the output of the **Score Model** component. As shown in the following screenshot, a new column has been added at the end of the dataset named `Scored Labels`. These are the predictions regarding the `Log GDP per capita` score for each row:

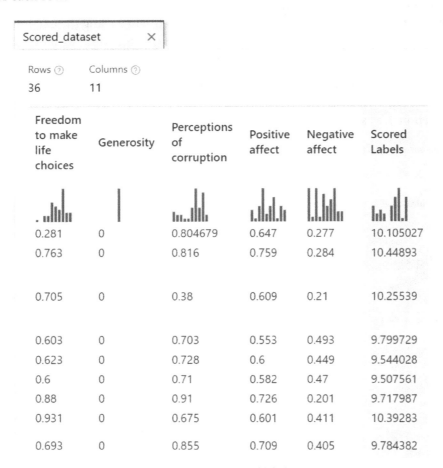

Figure 12.10 – Scored labels

Now that we know that the real-time inference pipeline works, it is time to deploy it so that we can integrate it with Power BI.

Creating the real-time endpoint

To deploy the model, you must create a real-time endpoint. A real-time endpoint typically uses a container-type compute to run the scoring script or inferencing pipeline.

To create a real-time endpoint using the Designer, follow these steps:

1. From the real-time inferencing pipeline in the Designer, select **Deploy** at the top right of the canvas. A new pane will open, as shown in the following screenshot:

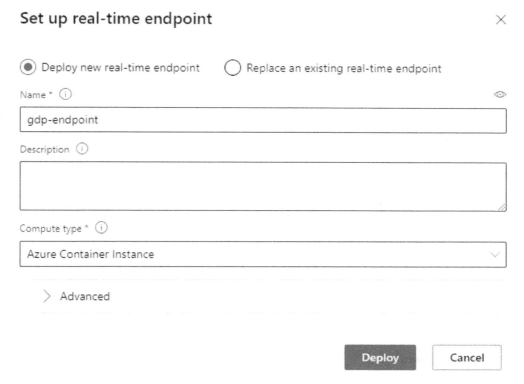

Figure 12.11 – Set up real-time endpoint

2. Create a new endpoint with the following settings:

    -   **Deploy new real-time endpoint**
    -   **Name:** gdp-endpoint
    -   **Compute type:** Azure Container Instance

Creating the endpoint and deploying the model to the endpoint will take around 10 minutes. Once the model has been deployed and the endpoint is ready, you'll get a notification. You can view its progress in the Studio by going to the **Endpoints** tab and selecting gdp-endpoint. The endpoint is ready to be used when **Deployment state** is **Healthy**, as shown in the following screenshot:

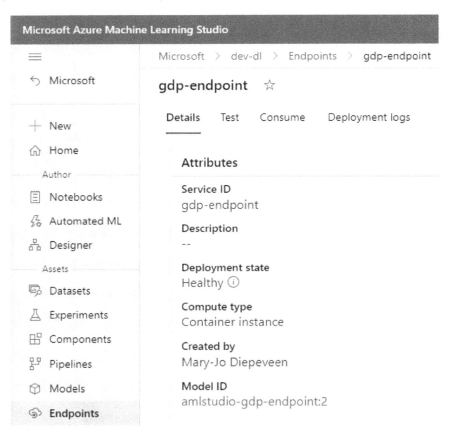

Figure 12.12 – Endpoint overview

Alternatively, when working with the *Python SDK* or *Azure CLI extension for Azure ML*, you can create a scoring script instead of using the inferencing pipeline. The scoring script should contain the same steps: process the data, load the model, and use it to generate new predictions. Using the Python SDK or the Azure CLI, you can create an endpoint and deploy the model. As a result, the endpoint will be listed in the *Studio* on the **Endpoints** tab.

With that, we have used the Designer to create a real-time inferencing pipeline and deployed it to an endpoint. The endpoint is healthy and ready to go. This means that we can finally integrate the model with Power BI to get predictions on data that is stored in Power BI.

Integrating an endpoint with Power BI to generate predictions

The final step of using Azure ML to train and deploy models is integrating the model. The purpose of training a model is often its consumption. Power BI's integration with Azure ML offers us that consumption without us having to set up complicated HTTP requests through Power Query Editor.

Assuming that you have a trained model in Azure ML and that you have deployed it to a real-time endpoint, you should be able to integrate that model with Power BI. To use that endpoint in Power BI Desktop, you have to sign in with the organizational account that also has access to the endpoint in the Azure ML workspace.

After signing in and importing the data, you can invoke an Azure ML real-time endpoint by using the **Azure Machine Learning** feature in Power Query Editor to add a new column with predictions.

Let's use the world happiness dataset once again and see it in action:

1. Open **Power BI Desktop**.
2. Import the world-happiness.csv dataset, which you downloaded from https://github.com/PacktPublishing/Artificial-Intelligence-with-Power-BI/tree/main/Chapter12/world-happiness-report.csv.
3. Open **Power Query Editor**.
4. In the **Add Column** tab, select **Azure Machine Learning** from the **AI Insights** section.

A pop-up will appear that shows all the available real-time endpoints associated with your account, as shown in the following screenshot:

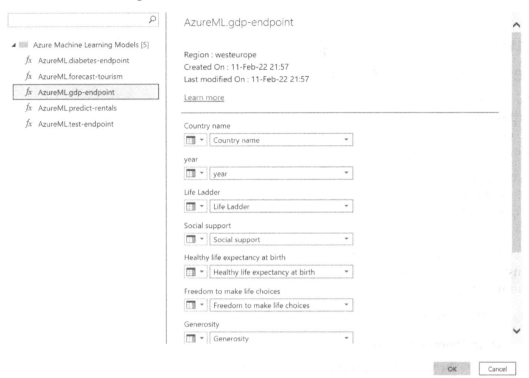

Figure 12.13 – Choosing an endpoint

5. Select `AzureML.gdp-endpoint` and ensure that the input columns are associated with the right fields. Select **OK**.

A new column will be added named `AzureML.gdp-endpoint`, as shown in the following screenshot:

Figure 12.14 – New column with predictions

The predictions are currently stored as records for the column name (`Stored Labels`) and the value is included for each row.

6. Expand the `AzureML.gdp-endpiont` column.

7. Ensure the `Scored Labels` column is selected.

8. Uncheck the **Use original column name as prefix** box.

Now, you should have a `Scored Labels` column with all the predictions for all the rows. In this use case, we are only using the predictions for data that was missing. Additionally, you can replace those values by creating a new conditional column.

9. Add a new **Conditional Column** with the following settings:

    - **New column name**: `Log GDP per capita`

    - **If Column Name**: `Actual GDP`

    - **Operator**: `does not equal`

    - **Value**: `null`

    - **Then Output**: (Set to **Select a column**) `Actual GDP`

    - **Else**: (Set to **Select a column**) `Scored Labels`

Finally, we have a `Log GDP per capita` column with no empty rows, as shown in the following screenshot:

1.2 Actual GDP	▼	ABC 123 Scored Labels	▼	ABC 123 Log GDP per capita	▼
● Valid	99%	● Valid	100%	● Valid	100%
● Error	0%	● Error	0%	● Error	0%
● Empty	< 1%	● Empty	0%	● Empty	0%
	7.37		7.224371732		7.37
	7.54		7.151323497		7.54
	7.647		7.410792291		7.647
	7.62		7.41407752		7.62

Figure 12.15 – New Log GDP per capita column

If you want to add predictions to your Power BI dataset that are not helping you fix missing data, you probably want to keep the `Scored Labels` column as-is or rename it to whatever the predictions entail.

To integrate a model with Power BI, you can take the trained model in Azure ML and deploy it to a real-time endpoint. The **Azure Machine Learning** feature in Power Query Editor will then allow you to add the predictions by adding a new column with the scored labels.

Summary

In this chapter, we explored Azure ML and how it can help us train and deploy models so that we can integrate them with Power BI. Throughout this chapter, we used the Azure ML Designer and Azure ML Studio to explore the process through the UI. All these steps can be replicated using the Python SDK or the Azure CLI extension for Azure ML if you want to use your training scripts and prefer to work with code. To get predictions in Power BI, you can choose to generate batch predictions upstream or deploy your model to a real-time endpoint to use the **Azure Machine Learning** feature in Power Query Editor. In the next chapter, we'll go over some things you should consider when you're training and deploying models so that you can use them responsibly.

13
Responsible AI

Throughout this book, we have explored ways to integrate **artificial intelligence (AI)** with Power BI, whether it is by using out-of-the-box features or by training your models. While AI can help uncover interesting insights from data, it is important to think about how the data and models are created and used. Insights are often used to make decisions and will influence your actions. Therefore, you want to be able to trust those insights and know that you are making responsible decisions.

There are many things to examine when practicing responsible AI. In this chapter, we'll discuss what responsible AI entails and three of the most common considerations. To do this, we will be covering the following topics:

- Understanding responsible AI
- Protecting privacy when using personal data
- Creating transparent models
- Creating fair models

Putting responsible AI into practice is a complicated endeavor that requires an interdisciplinary approach. We'll start by examining responsible AI before exploring the three most common concepts: privacy, transparency, and fairness.

Understanding responsible AI

Using insights that have been extracted from data to drive decisions is not necessarily a new practice. However, recent developments in society have raised many questions on what should or shouldn't be allowed when it comes to implementing AI. As a result, many organizations have started talking about how to use AI more responsibly.

So, what is meant by responsible AI? The idea of using AI responsibly mostly means that *AI should cause no harm to anyone*. There may be several ways in which we can cause harm. As AI is used to make decisions, we often look at how these decisions impact any individual. If we treat certain individuals unfairly or make the wrong decisions, we can cause them harm. Throughout this chapter, we'll explore how privacy, interpretability, and fairness are ways in which we can use AI responsibly.

Although these three concepts are discussed commonly in the context of responsible AI, they are not the only considerations for when you work with AI. To govern the use of AI within an organization, it is best practice to use an interdisciplinary approach. Different departments and experts should come together to work on responsible AI strategically and systematically.

In the context of business, this often means that guidelines need to be defined within an organization on when to work with data and how to handle models. People should also know who to go to when they feel that a current project or AI application isn't giving them advice on how to classify the application and how to mitigate any issues that may arise.

Many tools are provided, both open source and proprietary, to help achieve responsible AI. However, just like any tool, its effectiveness depends on the people using it. An organization needs to think about how it wants to achieve responsible AI and what it considers responsible to understand when to use which tool.

This chapter will serve as an introduction to responsible AI, what dilemmas may arise when working with AI, and how tools may help mitigate issues. Note that this chapter is not an exhaustive overview of useful resources or guidelines and should not be treated as such.

Instead, it focuses on the most common areas when talking about responsible AI:

- **Protecting privacy**: How to ensure someone's identity isn't compromised

- **Creating transparency**: How to ensure a model's prediction can be explained

- **Ensuring fairness**: How to ensure a model is equally accurate across groups

We'll delve into each of these common areas in the following sections.

Protecting privacy when using personal data

Whether a model needs any personal data as input depends on the use case. Similarly, it depends on what the model will be used for and what kind of personal data is needed. Let's start by going over what personal data can entail, and then how we can protect privacy when handling personal data.

Removing personally identifiable information

Personal data is any data that contains information about an individual. Most importantly, you need to know when you're working with **personally identifiable information** (**PII**) data. PII data is information that can be used to identify a specific individual – for example, someone's name, address, telephone number, or email address.

In general, PII data should be handled carefully. Even when it's needed for any processing you need to do (for example, to send someone a product they bought), people that don't need that information shouldn't be able to access it.

When it comes to AI, PII data is considered a *high cardinality feature*. A high cardinality feature is a field that contains many unique rows. When you're training a model, you want to find a pattern in the data. With high cardinality features, a pattern is difficult to observe. It could be that there is some related information to the PII data that does provide a model with information on what to predict.

For example, to predict whether someone is likely to buy a product from your company again, it is important to know where a customer lives. However, you don't need their exact address; the information the model needs will likely specify how close a customer lives to a store. Instead of using the PII data in cases like these, it is better to convert the data into an overarching feature, such as the distance from someone's home to the nearest store.

By converting personal data into an overarching feature, not only do you protect the privacy of individuals, but you also make sure that you understand the information the model is using to predict a target value.

If you don't convert someone's home address into the distance from their home to the nearest store, the model may still consider the home address as valuable information. However, you won't know whether it is because of some pattern based on proximity to stores or, for example, on whether a customer lives in a town or city.

It is important to know as much as you can about how your model reaches a decision, which is another reason why, instead of personal data, a data scientist should analyze what the overarching feature is that correlates with the target value.

Using differential privacy on personal data

Some personal data is off-limits. Whenever it's data that directly refers to an individual, such as PII data, you'll want to exclude it from your analysis. However, there may still be data about a person that is not considered PII data that can be used to trace them.

For example, let's say that you're analyzing the performance of students at a high school. You use their grades and exclude any PII data. However, there is still some demographic information available in the dataset, such as whether a student identifies as male or female and which class they're in. If someone were to gain access to that data, they could use this grades and gender information to trace it back to a specific individual (because only Daniel has a 4.2 GPA in mathematics).

The more personal data you include in your dataset, the easier it can be for someone to trace a data point to an individual. One solution to this problem can be to remove the personal data, by either minimizing the features that contain personal data or excluding any personal data. Nevertheless, models can benefit from this data, and sometimes, you'll want to include it to create a more accurate model.

Another solution is to apply **differential privacy**, an approach that *adds statistical noise to data to make it more private*. One tool that you can use to apply differential privacy to your data is the Python **SmartNoise** library (`https://smartnoise.org/`).

With a library like SmartNoise, you can tweak the differential privacy function to decide how much noise should be added to your data. The more noise you add, the more private the dataset becomes, which means the harder it is to trace information back to a certain individual. Though a library like SmartNoise will try to preserve the distribution of the data, adding noise may make your model more inaccurate too. It is through trial and error that you'll find the appropriate balance between privacy and accuracy.

Any personal data, whether personally identifiable or not, should be flagged in your dataset. How you handle any personal data will depend on what you need as input for training a model. You can either remove personal data, replace it with an overarching feature that extracts the necessary information, or use differential privacy to add statistical noise to the dataset.

Creating transparent models

One of the reasons AI has become increasingly popular over the past decade is the development of more complicated models, such as deep learning models. Deep learning models are especially successful on unstructured data as they can derive what features are needed to generate the prediction.

The benefit of deep learning models is that they are often more accurate, while the disadvantage is that they tend not to be very **transparent**: *it is unclear how the model generates a prediction or makes a decision.*

Using algorithms that are transparent by design

Transparency is becoming an increasingly important concern when it comes to training machine learning models. Even though more complicated algorithms can be used to train more accurate models, sometimes, a data scientist may opt for a simpler algorithm that is more transparent. The following diagram shows how simpler algorithms, such as linear models and decision trees, have better transparency but may produce less accurate models than neural networks, which are less transparent:

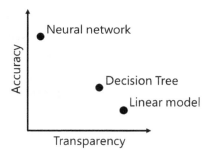

Figure 13.1 – Trade-off between transparency and accuracy

Whether you require your models to be transparent depends on the use case. If your model influences a person directly, you probably want – and need – to have a transparent model. Many financial institutions are expected to explain how any decision is made.

For example, if someone applies for a mortgage at a bank, and a model is used to predict the financial risk of the applicant, which is used to decide whether to provide that mortgage, the bank needs to be able to explain how the model predicted the risk. Especially when a mortgage gets rejected, a customer should be allowed to understand why. This will also help you identify any biases your model may have.

Simpler algorithms such as linear models and decision trees are transparent by design. For example, when you're training a model with a decision tree algorithm, you can obtain a decision tree, as shown in the following diagram:

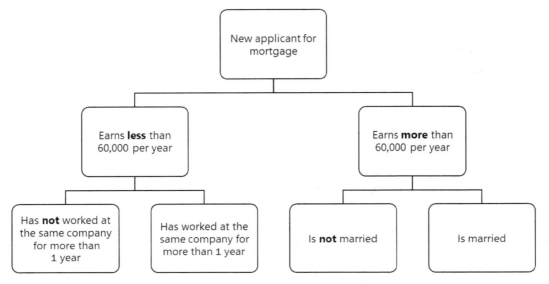

Figure 13.2 – Decision tree mortgage application

You can easily retrace how a model generates a prediction when using a transparent algorithm such as a decision tree when someone applies for a mortgage. The included features can be salary, time with current company, and marital status. The transparency of a model allows you to validate whether the appropriate decision has been made, but it also serves as a check on any bias your model may have.

For example, if the decision tree for the mortgage application shows marital status to be the first deciding feature, you may want to tweak your model. If we think about the reason why marital status appears important for classifying a customer as high or low risk, it could be because there is a financial backup (the partner of the applicant) for the customer. However, you may not want a model that decides to give someone a mortgage based primarily on their marital status. In this case, the model's transparency brings an unwanted bias to light.

Explaining black-box models

So, we prefer models to be transparent, especially in certain use cases, such as when the model is used for financial institutions or health care. But what if we want to use a more complicated algorithm that is not transparent by design as it provides more accurate predictions? When working with so-called **black-box** algorithms, *which are not transparent by design*, you can explain the trained models by using *interpretability techniques*.

With black-box models, an interpretability algorithm is used to approximate how important each feature is when predicting the target value. Different algorithms calculate the explanations with varying approaches. Which interpretability approach you use depends on the algorithm that's chosen during model training. Let's look at what such explanations may look like. Then, we'll delve into the techniques that can be used to calculate those approximate explanations.

In *Chapter 12, Training a Model with Azure Machine Learning*, we learned how to train a model with the **Azure Machine Learning** (**ML**) **Designer**. Throughout this book, we have worked with the world happiness dataset, which includes `Life Ladder`, a measure of happiness per country that's assessed on a scale of 1 to 10. Other factors that potentially influence the happiness of a country's citizens are included, such as `GDP per capita` and `healthy life expectancy at birth`, as shown in the following table:

Country name	year	Life Ladder	Log GDP per capita	Social support	Healthy life expectancy at birth	Freedom to make life choices	Generosity	Perceptions of corruption	Positive affect	Negative affect
Afghanistan	2018	2.694	7.692	0.508	52.6	0.374	-0.094	0.928	0.424	0.405
Afghanistan	2019	2.375	7.697	0.42	52.4	0.394	-0.108	0.924	0.351	0.502
Albania	2018	5.004	9.518	0.684	68.7	0.824	0.009	0.899	0.713	0.319
Albania	2019	4.995	9.544	0.686	69	0.777	-0.099	0.914	0.681	0.274

Figure 13.3 – World happiness data

Let's say we trained a regression model that can predict the `Life Ladder` score of a country based on the other fields as features. When you're training a model with the Designer, you have the option to set **Model explanations** to **True** to calculate model interpretability, as shown in the following screenshot:

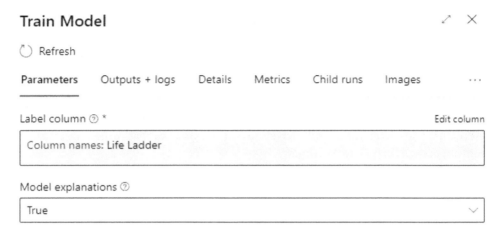

Figure 13.4 – Setting Model explanations to True

If you train a model using your training scripts with the SDK or CLI, you can use the `azureml-interpret` package, which has been developed by the Interpret Community (`https://github.com/interpretml/interpret-community/`).

Whatever approach you take to initiate model interpretability during model training, the explanations will be stored with the experiment run. You can review the explanations in **Azure ML Studio**, under the **Explanations** tab in the experiment run. For example, after training a model to predict the `Life Ladder` score for a country, you may get explanations similar to the following:

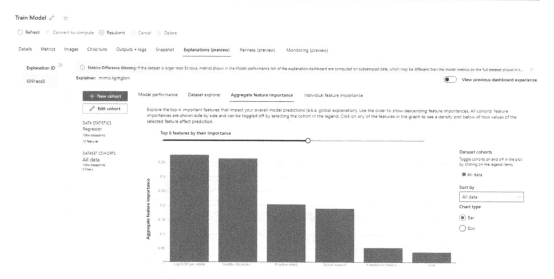

Figure 13.5 – Feature importance for log GDP per capita

Most importantly, as shown in the preceding screenshot, there are two tabs that you can explore:

- **Aggregate feature importance**: *The aggregate of the relative importance of each feature on predictions made on the test dataset.* This is mostly useful for data scientists to gain insights into how the model is working. It is often used as a sanity check to verify whether the model is biased or not.

- **Individual feature importance** (the chart will look similar to the one shown in the preceding screenshot): *The relative importance of each feature for predictions made on individual data points.* This becomes especially useful if you want to explain individual predictions when the model is deployed, such as when an individual customer asks you why the model predicted they should not get a mortgage.

For the trained model to predict the `Life Ladder` score, `GDP per capita` and `healthy life expectancy at birth` are the most important factors that contribute to a country's happiness score as the bars for those features are the highest.

So far, there's nothing to be concerned about. When you're working with demographic data, you will want to be aware of any bias that may exist in your data and model. For example, if marital status is the most important feature to predict a mortgage, then you probably want to revise your model and see how you can address this.

To generate feature importance, you can use different interpretability techniques. The three common approaches that are included in the `azureml-interpret` package are as follows:

- **SHAP**: This is based on game theory. It tries different combinations of features to see how they influence the prediction. The simulations are compared to the actual predictions to calculate the ranking of the feature's importance.

- **Mimic**: This trains a surrogate model. A simpler transparent algorithm is used on the same data with the same target value. The explanations are derived from the simpler model (as you would normally with a linear or decision tree model). The surrogate model is an approximation as it is expected to be less accurate than the black-box model (otherwise, you should use that simpler algorithm).

- **Feature permutation**: The model is retrained multiple times, each time taking a different combination of features to see how excluding features influences the model's accuracy. If a feature is excluded in a permutation and the model's accuracy is lower than that of the original model, it's an indication that the feature is important. Based on multiple permutations, the importance of each feature can be estimated.

> **Interpreting your model's explanations**
>
> Interpretability techniques try to approximate feature importance. Be careful with interpreting the model's explanations. How to interpret them will depend on the technique that's used. In general, you should draw conclusions based on the relative feature importance (the ranking of the feature importance) and not the absolute values of the calculated feature importance.

Although being able to calculate the model's interpretability allows us to get better insights into the workings of black-box models, you should always ask yourself whether it is worth trading off transparency for accuracy. When you're defining the use case, before training the model, you should decide on what level of transparency is required. Based on the transparency requirements, you may need to limit yourself to only certain algorithms when training a model.

So far, we have explored the importance of creating transparent models and discussed algorithms that may be more transparent by design and interpretability techniques that can be used to explain black-box models. Understanding the relative importance of features on a model's predictions may help identify bias. In the next section, we'll learn how to assess the fairness of models.

Creating fair models

Fairness is a complicated concept that is often used to reflect the equal treatment of different groups. We want a trained model to perform equally well across different groups. To ensure fairness, we need to assess whether a model is fair. If the model shows that it treats groups unfairly, we can retrain that model and force it to perform equally across groups.

Identifying unfairness in models

Imagine that we train a model to predict whether high school students will be successful when they continue their studies at a university. The model's predictions may influence a student's decision to apply to a university. As this is an important life decision, we want to make sure that the model predicts it correctly for both female and male students. We may also want to assess whether the model predicts it correctly for different minority groups.

Depending on the use case and the features you include when training a model, you may wish to identify so-called **sensitive features**: features that may divide your dataset that will assess that the model is fair.

Although these sensitive features may include personal or demographic data, you may still want to include them because they correlate with the target value. When you expect these kinds of features to influence the predictions, it is better to include them and analyze them for fairness, rather than to not include them at all. Often, we see that unfairness seeps through other features too, and not analyzing fairness across sensitive groups may result in you missing the unfairness instead of removing it.

For illustration purposes, let's think about the model we looked at previously, which predicts whether a high school student should go to university. Due to cultural trends, it may be that fewer women go to university and complete their courses. This may not have anything to do with their academic success.

As a result, the model may predict that men are likely to be more successful at university than women, which means the model may be biased in favor of men. A biased model isn't a problem by itself – it may be that this bias is fair for whatever reason. However, a bias is problematic when it turns out that the model is performing less well for the disadvantaged group.

If, for example, you wish to analyze and compare the model's accuracy (which percentage of the predictions are correct) for men and women, you will want the accuracy to be similar for each group. Whatever the predictions are, the accuracy should be equally high. If the accuracy turns out to be unequal between the groups, then it's an indication that the model is treating the groups unfairly.

To assess fairness between groups, you can analyze several evaluation metrics, such as accuracy and precision, depending on the algorithm that's used to train the model.

If unfairness is detected, we may say that there is a **disparity** in model performance. To force the model to treat the groups fairly, we can focus on a certain disparity that we want to avoid, which we can do with mitigating techniques.

Mitigating unfairness in models

Once you've decided what the sensitive features are and you've assessed whether a model performs equally well for each group, you may decide that the model is unfair and should be retrained or post-processed.

Assessing and mitigating unfairness can be done with an open source package such as **Fairlearn**. You can use Fairlearn with Azure ML or when you're training your models in another environment. It provides you with a dashboard to interactively inspect the sensitive features you've defined and assess whether the model is fair.

If unfairness is detected in a model, you can choose to either retrain the model with the restraint of ensuring the model performs well across groups, or you can post-process a trained model to minimize the unfairness.

For either approach, several algorithms exist that let you mitigate the unfairness in your model. Which approach you can use depends on how you trained the model and which algorithm you used. You can find an overview of the mitigation algorithms of Fairlearn here: https://docs.microsoft.com/en-us/azure/machine-learning/concept-fairness-ml#mitigation-algorithms.

Creating fair models can be a complicated task as disparity across groups may occur for a variety of reasons. Even though there are techniques you can use to assess fairness and mitigate it, it is a challenging task to create models that are fair across all groups. To ensure fairness, you should ensure your data is of good quality and identify sensitive features before training a model. After training, you can use the appropriate technique based on the algorithm that was used to train the model.

Summary

In this chapter, we explored the complicated concept of responsible AI. Implementing responsible AI requires an interdisciplinary team that has carefully thought about goals and guidelines. Common topics within responsible AI include protecting privacy when using personal data, creating transparent models, and ensuring your models are fair across groups.

Index

Other Books You May Enjoy

If you enjoyed this book, you may be interested in these other books by Packt:

Extending Power BI with Python and R

Luca Zavarella

ISBN: 9781801078207

- Discover best practices for using Python and R in Power BI products
- Use Python and R to perform complex data manipulations in Power BI
- Apply data anonymization and data pseudonymization in Power BI
- Log data and load large datasets in Power BI using Python and R

- Enrich your Power BI dashboards using external APIs and machine learning models
- Extract insights from your data using linear optimization and other algorithms
- Handle outliers and missing values for multivariate and time-series data
- Create any visualization, as complex as you want, using R scripts

Learn Power BI - Second Edition

Greg Deckler

ISBN: 9781801811958

- Get up and running quickly with Power BI
- Understand and plan your business intelligence projects
- Connect to and transform data using Power Query
- Create data models optimized for analysis and reporting
- Perform simple and complex DAX calculations to enhance analysis
- Discover business insights and create professional reports
- Collaborate via Power BI dashboards, apps, goals, and scorecards
- Deploy and govern Power BI, including using deployment pipelines

Packt is searching for authors like you

If you're interested in becoming an author for Packt, please visit `authors.packtpub.com` and apply today. We have worked with thousands of developers and tech professionals, just like you, to help them share their insight with the global tech community. You can make a general application, apply for a specific hot topic that we are recruiting an author for, or submit your own idea.

Share Your Thoughts

Now you've finished *Artificial Intelligence with Power BI*, we'd love to hear your thoughts! Scan the QR code below to go straight to the Amazon review page for this book and share your feedback or leave a review on the site that you purchased it from.

https://packt.link/r/1-801-81463-5

Your review is important to us and the tech community and will help us make sure we're delivering excellent quality content.

www.ingramcontent.com/pod-product-compliance
Lightning Source LLC
Chambersburg PA
CBHW062056050326
40690CB00016B/3109